Ex-BR DIESELS IN IN[illegible]Y

Full details of those British Rail diesel locomotives below 1,000hp sold for industrial service and preservation - past and present - at home and abroad

by
Adrian Booth

HANDBOOK 8BRD
INDUSTRIAL RAILWAY SOCIETY 2019

8BRD

ISBN: 978 1 912995 01 1

Eighth Edition

26 Great Bank Road, Herringthorpe,
Rotherham, South Yorkshire, S65 3BT

www.irsociety.co.uk

Front cover photograph: 03066 is a working locomotive at Barrow Hill Engine Shed at Staveley and it can sometimes be seen moving other locomotives and items of rolling stock around the yard. It is seen between shunts on 5th April 2017. (Adrian Booth)

Produced for the IRS by Print Rite, Freeland, Witney, Oxon. 01993 881662

INTRODUCTION

The first edition of this book was published in 1972, being conceived and edited by the late Eric Tonks, then the President of the Industrial Railway Society. In the mid-1970s, at Eric's request, I took over his records and I have now maintained, updated and published them for approximately forty-five years, up to the present day. The second edition of BRD was published in 1981, the third in 1987, the fourth in 1991, the fifth in 2000, the sixth in 2007, and the seventh in 2011.

The contents of the first four books were fairly straightforward, simply listing ex-British Railways diesel shunters working at industrial premises, together with those which had passed from industrial use into preservation. The handful of shunters that went directly to preservation were also listed, as they were akin to their industrial cousins. The information was compiled using notes and observations provided by members of the Society. Since publication of the fourth edition, however, there were massive changes in the British railway industry. This led to much thought as regards the contents of the fifth, sixth and seventh issues, in particular the effects of privatisation of main line railways, for it could be argued that ALL diesel shunters owned by privatised companies should then be included. Such a listing would have been impractical, however, and outside the scope of this series. A decision was taken (when preparing 5BRD) to include only shunters which worked at the privatised companies' workshops, and which were hired by them. These policies were continued from the sixth edition onwards.

The last two decades have seen many mergers, takeovers, and restructuring of companies, together with a move towards adopting 'modern' style names. A decision has been made to show company names as used at the time locomotives were acquired, and not to record subsequent name changes. Many private individuals have purchased locomotives for preservation, and these are simply shown under the name of their home preservation site, as it would be difficult to ascertain, update, and show ownership of these locomotives. An exception to this rule is Harry Needle who owns a large number of industrial and ex-BR locomotives and who formed the 'Harry Needle Railroad Company'. His locomotives are often kept at preservation sites, and herein are indicated by (HNRC) immediately after the site name. In recent years there has been an increase in the number of companies which acquire and then hire out locomotives, in particular Class 08s. By the very nature of such business hirings can be to industrial works or to main line depots or freight yards, and a decision has been made to include all hirings herein. This volume also continues the longstanding policy of listing only locomotives rated at up to 999hp.

I would like to thank: Alex Betteney, Brian Cuttell and Bob Darvill (regular BRD correspondents over many years); Ivor Thomas (re Adams hire locomotives); Cliff Shepherd (re Thomson's); John Cartwright (research in old IRS letters); Robert Pritchard (who suggested the new Appendix E and supplied the initial list); Peter Hall (miscellaneous queries, plus considerable help re exported locomotives, Adams hire locomotives and Alstom Transport); and Andrew Smith and Geoff Cryer (book production). Listings incorporate all information received by me up to 31st July 2019.

Adrian Booth
Rotherham, 31st July 2019
email : brlocos@irsociety.co.uk

EXPLANATORY NOTES

Layout

The book is divided into 28 sections, each devoted to one class, and brief details appropriate to each are given. The engines, etc, apply to locomotives as built, and not necessarily as when sold out of service. All are standard gauge unless otherwise stated. Sections 1 to 28 are followed by Appendices A to E. Each locomotive has eight columns of information, thus:

BR number	*Builder*	*Works number*	*Year built*	*Last depot*	*Date wdn*	*P/F*	*Title*

These will be simple to understand, but the following notes are pertinent:-

BR number

The 1957 numbering scheme (with the 'D' prefix) has been used to dictate the order of the classes in the book, although the later TOPS computer numbers (where allocated) are also given.

Builder

Many locomotives were built at British Railways' own workshops at Ashford, Crewe, Darlington, Derby, Doncaster, Horwich and Swindon, and these are shown where appropriate. Locomotives which were constructed at private companies' workshops are indicated by standard IRS abbreviations, as under. In the case of Drewry locomotives two numbers are given: this company was basically a sales organisation, which did not erect locomotives but rather sub-contracted the construction work, in practice to RSH or VF. Both organisations' works numbers are given in these cases.

AB	Andrew Barclay, Sons & Co Ltd, Kilmarnock
CE	Clayton Equipment Co Ltd, Hatton, Derby
DC	Drewry Car Co Ltd, London
EE	English Electric Co Ltd, London
HC	Hudswell, Clarke & Co Ltd, Leeds
HE	Hunslet Engine Co Ltd, Leeds
NB	North British Locomotive Co Ltd, Glasgow
RH	Ruston & Hornsby Ltd, Lincoln
RSHD	*Robert Stephenson & Hawthorns Ltd, Darlington
RSHN	*Robert Stephenson & Hawthorns Ltd, Newcastle upon Tyne
VF	*Vulcan Foundry Ltd, Newton-le-Willows, Lancashire
YE	Yorkshire Engine Co Ltd, Sheffield

*(latterly owned by English Electric)

Works number

This only applies to locomotives built by private companies. Those locomotives constructed at British Railways workshops were not allocated individual works numbers.

Year built

The year quoted is that in which the locomotive was officially added to British Railways stock, or the worksplate date if this is known to be different.

Last depot

During their 'main line' life, all the locomotives listed in this book were allocated to various British Railways motive power depots. Up to 6th May 1973 these depots were identified by an official code that comprised a number and letter; after this date a new system of two-letter codes was adopted. The list below shows all the depots and workshops from which locomotives listed in this book were withdrawn. The original and new codes for all of these depots are shown below although, in most cases, the older style code is used in the listings.

1A	WN	Willesden (London)
1E	BY	Bletchley
2E	SY	Saltley (Birmingham)
2F	BS	Bescot (Walsall)
5A	CD	Crewe Diesel
6A	CH	Chester
6G	LJ	Llandudno Junction
8C	—	Speke Junction (Liverpool)
8F	SP	Springs Branch (Wigan)
8H	BC	Mollington Street (Birkenhead)
	BD	Birkenhead North
8J	AN	Allerton (Liverpool)
9A	LO	Longsight (Manchester)
9D	NH	Newton Heath (Manchester)
10D	LH	Lostock Hall (Preston)
12A	KD	Kingmoor Diesel (Carlisle)
12B	CL	Upperby (Carlisle)
12C	BW	Barrow
15A	LR	Leicester Midland
16A	TO	Toton (Nottingham)
16B	—	Colwick (Nottingham)
16C	DY	Derby
16F	BU	Burton upon Trent
30A	SF	Stratford (London)
	TM	Temple Mills (sub of 30A)
30E	CR	Colchester
31A	CA	Cambridge
31B	MR	March
32A	NR	Norwich Thorpe
	NC	Norwich Crown Point
34E	PB	New England (Peterborough)
36A	DR	Doncaster
36C	FH	Frodingham (Scunthorpe)
40A	LN	Lincoln
40B	IM	Immingham
41A	TI	Tinsley (Sheffield)
41J	SB	Shirebrook West
50B	—	Dairycoates (Hull)
50C	BG	Botanic Gardens (Hull)
50D	GO	Goole
51A	DN	Darlington
51L	TE	Thornaby (Middlesbrough)
52A	GD	Gateshead
52B	HT	Heaton (Newcastle)
—	TY	Tyne Yard (Gateshead)
55A	HO	Holbeck (Leeds)
55B	YK	York
—	YC	York Clifton (from Oct. 1983)
55C	HM	Healey Mills (Wakefield)
55F	HS	Hammerton Street (Bradford)
55G	KY	Knottingley
55H	NL	Neville Hill (Leeds)
60A	IS	Inverness
62A	TJ	Thornton Junction (Kirkcaldy)
62C	DT	Townhill (Dunfermline)
64B	HA	Haymarket (Edinburgh)
64H	—	Leith Central (Edinburgh)
—	EC	Craigentinny (Edinburgh)
65A	ED	Eastfield (Glasgow)

66A	PO	Polmadie (Glasgow)
66B	ML	Motherwell
67C	AY	Ayr
70D	EH	Eastleigh
70F	BM	Bournemouth West
70H	RY	Ryde (Isle of Wight)
73C	HG	Hither Green (London)
73F	AF	Chart Leacon (Ashford)
75A	BI	Brighton
75C	SU	Selhurst (London)
81A	OC	Old Oak Common (London)
81D	RG	Reading
81F	OX	Oxford
82A	BR	Bath Road (Bristol)
82B	PM	St Philip's Marsh (Bristol)
82C	SW	Swindon
82D	WY	Westbury
83B	—	Taunton
84A	LA	Laira (Plymouth)
85A	WS	Worcester
85B	GL	Gloucester
86A	CF	Canton (Cardiff)
86E	ST	Severn Tunnel Junction
87B	MG	Margam
87E	LE	Landore (Swansea)
	ZC	Crewe Works
	ZF	Doncaster Works
	ZG	Eastleigh Works
	ZH	St Rollox (Glasgow) Works
	ZJ	Horwich Works
	ZI	Ilford Works
	ZL	Swindon Works
	ZN	Wolverton Works

The following are codes that have been adopted for this book only:-

BSD	Beeston Sleeper Depot, near Nottingham
CJ	Chesterton Junction Permanent Way Materials Depot, Cambridge
HHC	Hall Hills Creosoting Depot, Boston
LSD	Lowestoft Sleeper Depot, Suffolk
MQ	Meldon Quarry, near Okehampton, Devon
RSD	Reading Signal Depot

Date withdrawn

This is the official date of withdrawal from British Railways stock. Official BR information often quoted a locomotive being withdrawn during a period, as for example "7-12-86 to 3-1-87". In such cases the last date has been taken, and the example quoted would be shown as 1/87 in the following tables. As regards post-BR days, policy changed so that locomotives tended not to be allocated to specific depots, whilst the TOCs (Train Operating Companies) tended not to release official information regarding withdrawals. It has thus proved difficult to find information as regards withdrawal depots and dates, resulting in many locomotives not having this information shown.

P/F and Title

These columns indicate whether the stated number and/or name is the Present (P) or Former (F) one carried by the locomotive. NPT indicates 'No Present Title', meaning the locomotive presently carries neither name nor number. Former titles given for scrapped locomotives are those carried by the locomotive at the time it was cut up. NFT indicates 'No former title', meaning the locomotive carried neither name nor number at the time it was scrapped.

Notes

These follow on from the basic locomotive data. The first industrial user (or preservation society) is stated, together with the date the locomotive arrived on site. Locomotives were probably moved from their last main line depot unless otherwise stated, and in some cases despatch details are shown where positive information has been found. All known subsequent movements and ultimate disposals are given, with the dates when known. Certain abbreviations are used:

APCM	Associated Portland Cement Manufacturers Ltd
BHESS	Barrow Hill Engine Shed Society
BR	British Railways
BSC	British Steel Corporation
CEGB	Central Electricity Generating Board
DBS	Deutsch Bahn Schenker
ECC	English China Clay
EWS	English Welsh & Scottish Railway Ltd
GBRf	GB Railfreight Ltd
GNER	Great North Eastern Railway Ltd
HNRC	Harry Needle Railroad Company
NCB	National Coal Board
NCBOE	National Coal Board Opencast Executive
NSF	National Smokeless Fuels Ltd
SYRPS	South Yorkshire Railway Preservation Society

LOCOMOTIVE LISTINGS

BR number	*Builder*	*Works number*	*Year built*	*Last depot*	*Date wdn*	*P/F*	*Title*

SECTION 1

British Railways built 0-6-0 diesel mechanical locomotives, numbered D2000-D2199, and D2370-D2399, introduced 1957. Fitted with a Gardner 8L3 engine developing 204bhp at 1200rpm, a five speed gearbox, and driving wheels of 3ft 7in diameter. Later classified TOPS Class 03.

D2012 Swindon 1958 31B 12/75 F 03012 / F135L
03012 to A. King & Sons Ltd, Snailwell, Cambridgeshire, 28th July 1976; scrapped 28/29th January 1991.

D2018 Swindon 1958 32A 11/75 P No.2 / 600
03018 to George Cohen, Sons & Co Ltd, Cransley, Northamptonshire, 29th April 1976; to 600 Fragmentisers Ltd, Willesden, London, October 1980; to Mayer Parry Ltd, Snailwell, Cambridgeshire, 17th July 1995; to South Yorkshire Railway Preservation Society (HNRC), Meadowhall, Sheffield, 11th November 1998; to Lavender Line, Isfield, about June 2001; to Mangapps Farm Railway Museum, Burnham-on-Crouch, 1st March 2004.

D2020 Swindon 1958 32A 12/75 P 03020 / F134L
03020 to A. King & Sons Ltd, Snailwell, Cambridgeshire, July 1976; to South Yorkshire Railway Preservation Society (HNRC), Meadowhall, Sheffield, 9th November 1995; to Lavender Line, Isfield, for storage, 27th July 2001; to Sonic Rail Services Ltd, East Newlands, St Lawrence, Southminster, Essex, 2012.

D2022 Swindon 1958 52A 11/82 P 2022
03022 to Swindon & Cricklade Railway; despatched from BR Swindon Works, 18th November 1983; to Coopers (Metals) Ltd, Swindon, on hire, early October 1989; returned to Swindon & Cricklade Railway, summer 1995.

D2023 Swindon 1958 40A 7/71 P D2023
to Tees & Hartlepool Port Authority, Middlesbrough Docks, July 1972; to T&HPA, Grangetown Docks, 15th September 1980; to Kent & East Sussex Railway, Tenterden, 14th August 1983.

D2024 Swindon 1958 40A 7/71 P D2024
to Tees & Hartlepool Port Authority, Middlesbrough Docks, July 1972; to T&HPA, Grangetown Docks, 15th September 1980; to Kent & East Sussex Railway, Tenterden, 4th September 1983.

D2027 Swindon 1958 30E 1/76 P NPT
03027 to Shipbreaking (Queenborough) Ltd, Kent, July 1976; to South Yorkshire Railway Preservation Society, Meadowhall, Sheffield, 9th February 1991; to Knights of Old Ltd, Old, Northamptonshire, 28th September 1993; to Peak Rail, Darley Dale, 2nd January 1997.

D2037 Swindon 1959 32A 9/76 P NPT

03037 to Hargreaves Industrial Services Ltd, NCBOE British Oak Disposal Point, Crigglestone; despatched from 32A Crown Point Depot, Norwich, October 1977; to NCBOE West Hallam Disposal Point, Mapperley, November 1983; to NCBOE British Oak Disposal Point, Crigglestone, July 1984; to NCBOE Oxcroft Disposal Point, Clowne, November 1988; to South Yorkshire Railway Preservation Society (HNRC), Meadowhall, Sheffield, 18th July 1995; to Lavender Line, Isfield, about June 2001; to Peak Rail, Rowsley, 18th May 2004; to private owner, near Burnham, Somerset, early 2012; to Foxfield Railway, Staffordshire, about 6th July 2012; to Royal Deeside Railway, Banchory, 16th October 2012.

D2041 Swindon 1959 75C 2/70 P D2041

to CEGB Richborough Power Station, February 1970; to CEGB Rye House Power Station, Hoddesdon, Hertfordshire, March 1971; to CEGB Barking Power Station, about May 1971; to CEGB Rye House Power Station, Hoddesdon, August 1974; to Colne Valley Railway, Castle Hedingham, Essex, 15th January 1981.

D2046 Doncaster 1958 51L 10/71 P NPT

Rebuilt by Hunslet (6644 of 1967) including flameproof equipment; to Gulf Oil Co Ltd, Waterston, Milford Haven, May 1972; to BR Canton Depot, Cardiff, for tyre turning, 16th May 1975; returned to Gulf Oil Co Ltd, Waterston; to BR Canton Depot, Cardiff, for tyre turning, 5th November 1980; returned to Gulf Oil Co Ltd, Waterston; to Petro-Plus, Waterston, Milford Haven, with site, 1998; to Beavor Locomotive Company, Dowlais, Merthyr Tydfil, January 2005; to Moveright International, Wishaw, Warwickshire, about March 2006; to Plym Valley Railway, Marsh Mills, Plymouth, 28th May 2008.

D2049 Doncaster 1958 50D 8/71 F D2049

to Hargreaves Industrial Services Ltd, NCBOE Bowers Row Disposal Point, Astley, West Yorkshire; despatched from 50D Goole Depot, by 12th May 1974; to NCBOE British Oak Disposal Point, Crigglestone, by 6th January 1975; to NCBOE West Hallam Disposal Point, Mapperley, Derbyshire, March 1978; to NCBOE British Oak Disposal Point, Crigglestone, 28th January 1985; scrapped on site by Wath Skip Hire Ltd of Rotherham, November 1985.

D2051 Doncaster 1959 30E 12/72 P D2051

to Ford Motor Co Ltd, Dagenham; despatched from 30E Colchester Depot, June 1973; to BR Swindon Works, for rebuild, 3rd October 1977; to Ford Motor Co Ltd, Dagenham, 23rd February 1978; to Rother Valley Railway, Robertsbridge, East Sussex, 6th November 1997; to North Norfolk Railway, Sheringham, by 28th September 2000.

D2054 Doncaster 1959 55B 11/72 F CENTA

to Chair Centre Ltd, Derby, October 1973; to British Industrial Sand Ltd, Middleton Towers, Norfolk, 20th July 1979; to C.F. Booth Ltd, Rotherham, May 1982; scrapped, September 1982.

D2057 Doncaster 1959 51L 10/71 F No.1

Rebuilt by Hunslet (6645 of 1967); to NCB Grimethorpe Colliery, Barnsley, 19th September 1972; to C.F. Booth Ltd, Rotherham, 25th March 1986; scrapped, April 1986.

D2059 **Doncaster** **1959** **NC** **7/87** **P** **D2059 / EDWARD**
03059 to Isle of Wight Steam Railway, Havenstreet; despatched from BR Colchester Depot, 4th November 1988; to Island Line, on hire, 16th March 2002; returned to Isle of Wight Steam Railway, Havenstreet, 19th March 2002.

D2062 **Doncaster** **1959** **32A** **12/80** **P** **D2062**
03062 to Dean Forest Railway, Lydney; despatched from BR Swindon Works, 30th September 1982; to Yeovil Country Railway, Yeovil Junction, about September 1995; to East Lancashire Railway, Bury, 13th October 1997; to Metrolink, Manchester, on hire, (upgrade contract), 26th May 2007; returned to East Lancashire Railway, Bury, about 12th September 2007.

D2063 **Doncaster** **1959** **52A** **12/87** **P** **D2063**
03063 to Colne Valley Railway, Castle Hedingham, 11th November 1988; to East Anglian Railway Museum, Wakes Colne, Essex, March 1995; to North Norfolk Railway, Sheringham, 23rd February 2000.

D2066 **Doncaster** **1959** **52A** **1/88** **P** **03066**
03066 to Horwich Foundry Ltd, Horwich; despatched from BR Gateshead Depot, 28th June 1988; arrived at Horwich, 30th June 1988; to South Yorkshire Railway Preservation Society, Meadowhall, Sheffield, 5th April 1991; to Barrow Hill Engine Shed Society, Staveley, 23rd March 1999.

D2069 **Doncaster** **1959** **52A** **12/83** **P** **D2069**
03069 to Vic Berry Company, Leicester, 4th January 1984; auctioned by Walker Walter Hanson of Nottingham, 1st August 1991, and sold for £9,500; to the Gloucestershire Warwickshire Railway, Toddington, 2nd August 1991; to Wabtec, Doncaster (for open day), July 2003; returned to GWR Toddington, May 2004; to Crewe Works, for open day, September 2005; returned to GWR Toddington, September 2005; seen on low-loader at Strensham North, M6 motorway (destination not known), 6th January 2006; returned to GWR Toddington by 2nd April 2006; to Vale of Berkeley Railway, Sharpness, 7th December 2015.

D2070 **Doncaster** **1959** **51L** **11/71** **F** **D2070**
to Shipbreaking (Queenborough) Ltd, Kent, June 1972; seen on 11th July 1972; to South Yorkshire Railway Preservation Society, Meadowhall, Sheffield, 5th October 1990; to Churnet Valley Railway, Cheddleton, 11th September 1993; to Cotswold Rail, RAF Quedgeley, near Gloucester, March 2001; scrapped, July 2001.

D2072 **Doncaster** **1959** **51A** **3/81** **P** **D2072**
03072 to Lakeside & Haverthwaite Railway Co Ltd, Cumbria; despatched from 51A Darlington Depot, August 1981.

D2073 **Doncaster** **1959** **BD** **5/89** **P** **03073**
03073 to BR Chester for storage, 30th March 1989; despatched from BR Chester, 17th January 1991; arrived at Railway Age (later renamed Heritage Centre), Crewe, 18th February 1991; displayed at Crewe Depot open day, 12th October 1991; to East Lancashire Railway, Bury, 1st October 1992; returned to Railway Age, Crewe, 8th October 1992; displayed at Crewe Railfare, 21st August 1994; to Greater Manchester Metro Ltd,

on hire, 19th August 1995; returned to Railway Age, Crewe, 16th September 1995; to Greater Manchester Metro Ltd, on hire, 26th September 1996; returned to Railway Age, Crewe, 8th October 1996; to Greater Manchester Metro Ltd, on hire, 5th September 1997; returned to Railway Age, Crewe, 22nd September 1997; to Greater Manchester Metro Ltd, on hire, 12th March 1999; returned to Railway Age, Crewe, 30th March 1999; to Greater Manchester Metro Ltd, on hire, 9th January 2001; returned to Railway Age, Crewe, by 3rd March 2001; to Crewe Works, for open day, 3rd September 2005; returned to Railway Age, Crewe, early 2006.

D2078 Doncaster 1959 52A 1/88 P 03078
03078 to Stephenson Railway Museum, Chirton, near Newcastle upon Tyne, 11th May 1988.

D2079 Doncaster 1960 70H 6/98 P 03079
03079 sold by Island Line, Isle of Wight (BR's successor); to Derwent Valley Light Railway Society, Murton, York, 6th June 1998.

D2081 Doncaster 1960 31B 12/80 P 03081 / LUCIE
03081 ex Appendix C; to Mangapps Farm Railway Museum, Burnham-on-Crouch, 8th March 2004.

D2084 Doncaster 1959 NC 6/87 P D2084
03084 to Knights of Old Ltd, Old, Northamptonshire; despatched from 31B March Depot, 26th January 1992; to Peak Rail, Darley Dale, 4th January 1997; to Ecclesbourne Valley Railway, Wirksworth, 16th March 2005; to LH Group, Barton under Needwood, for repairs, 7th July 2006; returned to Ecclesbourne Valley Railway, Wirksworth, 21st March 2007; to Lincolnshire Wolds Railway, Ludborough, 15th June 2009; to West Coast Railway Company, Carnforth, 11th January 2011; to East Lancashire Railway, Bury, 3rd March 2016; stayed at Bury for gala held in mid-April 2016; to West Coast Railway Company, Carnforth, about May 2016.

D2089 Doncaster 1960 NC 12/87 P 03089
03089 to British Sugar Corporation Ltd, Woodston Factory, Peterborough, for storage; despatched from 31B March Depot, 2nd September 1988; to Nene Valley Railway, Wansford, about May 1991; to Mangapps Farm Railway Museum, Burnham-on-Crouch, 3rd October 1991.

D2090 Doncaster 1960 55B 7/76 P D2090 / VIN
03090 to National Railway Museum, York, August 1976; to National Railway Museum, Shildon, 10th June 2004; to Midland Road Depot, Leeds, for tyre turning, 13th November 2015; to National Railway Museum, Shildon, 23rd November 2015.

D2093 Doncaster 1960 51L 10/71 F No.2
Rebuilt by Hunslet (6643 of 1967); to NCB Grimethorpe Colliery, Barnsley, 19th September 1972; to C.F. Booth Ltd, Rotherham, 26th March 1986; scrapped, 21st April 1986.

D2094 Doncaster 1960 52A 1/88 P D2094
03094 to Horwich Foundry Ltd, Horwich; despatched from BR Gateshead Depot, 28th

June 1988; arrived at Horwich, 30th June 1988; to South Yorkshire Railway Preservation Society, Meadowhall, Sheffield, 6th April 1991; to Barrow Hill Engine Shed Society, Staveley, 23rd March 1999; to Cambrian Railway Trust, Llynclys, Oswestry, 2nd June 2005; to Royal Deeside Railway, Banchory, 13th May 2010.

D2099 Doncaster 1960 51L 2/76 P 03099

03099 to National Smokeless Fuels Ltd, Fishburn Coking Plant, County Durham, July 1976; to Monkton Coking Plant, Hebburn, 26th February 1981; to South Yorkshire Railway Preservation Society, Meadowhall, Sheffield, 23rd April 1992; to Exhibition of Steam and Speed, Doncaster, for display, July 2000; returned to SYRPS, 17th July 2000; to Peak Rail, Rowsley, March 2002.

D2112 Doncaster 1960 NC 7/87 P D2112

03112 to British Sugar Corporation Ltd, Woodston Factory, Peterborough, for storage; despatched from 31B March Depot, 2nd September 1988; to Nene Valley Railway, Wansford, about May 1991; to Port of Boston, on hire, 20th January 1992; returned to Nene Valley Railway, 29th January 1992; to Port of Boston, on hire, 11th September 1998; to Nene Valley Railway, Wansford, for display, 5th October 2006, to Port of Boston, on hire, 8th October 2006; purchased by Harry Needle Railroad Company, June 2011; to Kent & East Sussex Railway, Tenterden, 9th November 2011; to Rother Valley Railway, Robertsbridge, East Sussex, 13th July 2012.

D2113 Doncaster 1960 55B 8/75 P 03113

03113 to Gulf Oil Co Ltd, Waterston, Milford Haven, 17th May 1976; to Maritime & Heritage Museum, Milford Haven, 17th October 1991; to Peak Rail, Rowsley, 6th to 16th November 2002.

D2114 Swindon 1959 82C 5/68 F D2114

to Bird's Commercial Motors Ltd, Long Marston, Worcestershire; despatched from BR Swindon Depot, August 1968; scrapped, January 1975.

D2117 Swindon 1959 8F 10/71 P D2117

to Lakeside & Haverthwaite Railway Co Ltd, Cumbria; despatched from Springs Branch Depot, Wigan, and ran under own power to Ulverston, 14th April 1972; completed journey by road, 24th April 1972.

D2118 Swindon 1959 12C 6/72 P 03118

to Anglian Building Products Ltd, Atlas Works, Lenwade, Norfolk, August 1973; to Costain Dow-Mac Ltd, Tallington Works, Lincolnshire, 7th December 1993; to South Yorkshire Railway Preservation Society (HNRC), Meadowhall, Sheffield, 25th March 1996; to Rutland Railway Museum, Cottesmore, on loan, 20th March 2001; to Peak Rail, Rowsley, 31st March 2005; to Great Central Railway, Ruddington, Nottingham, 28th January 2011.

D2119 Swindon 1959 87E 2/86 P 03119

03119 to A.E. Knill Ltd, Barry, by 18th November 1986; to Dean Forest Railway, Lydney, 19th December 1986; to Rail & Maritime Engineering, Thingley Junction, Chippenham, for storage, about March 1995; to West Somerset Railway, Minehead, 28th March 1996; to Epping Ongar Railway, Essex, 3rd December 2011.

D2120 Swindon 1959 87E 2/86 P D2120

03120 to Sir W.H. McAlpine, The Fawley Hill Railway, near Henley-on-Thames, Buckinghamshire, 18th December 1986; to Didcot Railway Centre, Oxfordshire, for gala, 20th May 2015; returned to The Fawley Hill Railway, 28th May 2015.

D2122 Swindon 1959 82A 11/72 F D2122

to John Cashmore Ltd, Newport, 1974; to Briton Ferry Steel Co Ltd, Glamorgan, in a dismantled condition, 13th August 1974; used for spares; remains scrapped by August 1976.

D2123 Swindon 1959 82C 12/68 F NFT

to Bird's Commercial Motors Ltd, Long Marston, Worcestershire; despatched from 82C Swindon Depot, June 1969; to Bird's (Swansea) Ltd, 40 Acre Site, Cardiff, April 1970; converted to a stationary generator; seen at Cardiff, 20th August 1970; to Derby Power Station, Full Street, Derby, by 8th November 1970, where used as a generator on demolition contract; to Bird's, Stapleton Road, Bristol, for storage, 1971; later used as a stationary generator; seen there in a dismantled state, November 1978; scrapped shortly afterwards.

D2125 Swindon 1959 82C 12/68 F NFT

to Birds Commercial Motors Ltd, Long Marston, Worcestershire; despatched from 82C Swindon Depot, June 1969; to Bird's (Swansea) Ltd, 40 Acre Site, Cardiff, by 30th June 1969; seen at 40 Acre Site, 19th March 1972; seen re-painted in green livery with red chimney and no BR number, 20th December 1972; scrapped, June 1976.

D2128 Swindon 1960 82A 7/76 P 03901

03128 to Bird's Commercial Motors Ltd, Long Marston, October 1976; resold for export, minus its engine; shipped from Harwich Docks, December 1976; see Appendix C; returned to England and moved to Peak Rail, Darley Dale, 29th June 1993; to Nottingham Sleeper Co Ltd, Elkesley, Retford, 24th April 1996; to Cotswold Rail, Fire Service College, Moreton in Marsh, about September 2000; to Dean Forest Railway, Lydney, about July 2002; to Blast Furnace Sidings, Corus, Scunthorpe, for storage, 8th July 2008; to Appleby Frodingham Railway Society, Scunthorpe, August 2008; rebuilt as 0-6-0 diesel-hydraulic by Andrew Briddon, September 2011; to Andrew Briddon, Darley Dale, 8th September 2016; to Fox Productions Ltd, Longcross, Surrey, 16th December 2016; to Andrew Briddon, Darley Dale, 1st March 2017.

D2132 Swindon 1960 82C 5/69 F D2132 / LESLEY

to NCB Bestwood Colliery, Nottinghamshire, October 1970; to New Hucknall Colliery, Huthwaite, July 1971; to Pye Hill Colliery, Jacksdale, about 1981; to C.F. Booth Ltd, Rotherham, for scrap, 27th November 1984; scrapped, 28th/29th November 1984.

D2133 Swindon 1960 82A 7/69 P D2133

to British Cellophane Ltd, Bridgwater, Somerset, July 1969; to West Somerset Railway, Minehead, 10th July 1996.

D2134 Swindon 1960 82A 7/76 P D2134

03134 to Bird's Commercial Motors Ltd, Long Marston, October 1976; resold for export, about January 1977; see Appendix C; returned to England and moved to South

Yorkshire Railway Preservation Society, Meadowhall, Sheffield, 27th April 1995; to Royal Deeside Railway, Banchory, 12th December 2000.

D2138 Swindon 1960 82C 5/69 P D2138
to NCB Bestwood Colliery, Nottinghamshire, 23rd October 1970; to Pye Hill Colliery, Jacksdale, April 1971; to BR Toton Depot, for tyre turning, June 1978; returned to Pye Hill Colliery, Jacksdale, 19th June 1978; to Midland Railway, Butterley, Derbyshire, 20th August 1985.

D2139 Swindon 1960 85A 5/68 P D2139
to A.R. Adams & Son, Newport; despatched from 85A Worcester Depot, 10th December 1968; used as hire locomotive (see Appendix A); sold to NSF Coed Ely Coking Plant, Tonyrefail, by March 1971; to BR Canton Depot, Cardiff, for repairs, 29th September 1977; to Coed Ely Coking Plant, Tonyrefail, 1978 (after 18th March 1978); to BR Swindon Works, for repairs, 31st March 1981; to Coed Ely Coking Plant, Tonyrefail, 2nd March 1982; to Monkton Coking Plant, Hebburn, December 1983; to South Yorkshire Railway Preservation Society (HNRC), Meadowhall, Sheffield, 24th April 1992; to Peak Rail, Rowsley, March 2002.

D2141 Swindon 1960 87E 7/85 P 03141
03141 to White Wagtail Ltd, c/o DeMulder & Sons, Gun Range Farm, Shilton, near Coventry; despatched from 86E Severn Tunnel Junction Depot, 30th June 1986; to Cotswold Rail, Fire Service College, Moreton in Marsh, about December 2000; to Cotswold Rail, RAF Quedgeley, near Gloucester, March 2001; to Dean Forest Railway, Lydney, on hire, spring 2002; to Swansea Vale Railway, 19th June 2005; to Pontypool & Blaenavon Railway, 29th April 2008.

D2144 Swindon 1960 87E 2/86 P 03144
03144 to MoD Long Marston, Worcestershire, 24th March 1987; to MoD Grantham Barracks, for display, June 1992; to MoD Bicester, 26th June 1992; to Yorkshire Engine Company, Long Marston, 11th December 1995; to 275 Squadron, MoD Bicester, 5th September 1996; to Wensleydale Railway, Leeming Bar, by 25th January 2004.

D2145 Swindon 1960 87E 7/85 P 03145
03145 to White Wagtail Ltd, c/o DeMulder & Sons, Gun Range Farm, Shilton, near Coventry; despatched from BR Gloucester, about June 1986; to Cotswold Rail, Fire Service College, Moreton in Marsh, about December 2000; to D2578 Locomotive Group, Moreton on Lugg, 6th August 2001.

D2146 Swindon 1961 87E 9/68 F D2146
to Bird's Commercial Motors Ltd, Long Marston, Worcestershire; despatched from 82C Swindon Depot, June 1969; used in Army training exercise, MoD Long Marston, 16th October 1971; returned to Bird's, Long Marston; seen at Bird's, early November 1971 and 1st September 1972; scrapped, about August 1978.

D2148 Swindon 1960 55C 11/72 P D2148
to Hargreaves Industrial Services Ltd, NCBOE Bowers Row Disposal Point, Astley, Yorkshire; despatched from BR Healey Mills Depot, August 1973; to Lindley Plant Ltd,

NCBOE Gatewen Disposal Point, Denbighshire, September 1973; to NCBOE Bowers Row Disposal Point, December 1973; following collision damage to its original cab, 03149's cab was purchased from BR Doncaster Works and fitted on site, January 1984; to Steamport Transport Museum, Southport, 14th March 1987; to Ribble Steam Railway, Preston, 17th April 1999; to EWS Crewe Electric Depot, for tyre turning, 23rd January 2009; to Ribble Steam Railway, Preston, 15th February 2009.

D2150 Swindon 1960 55B 11/72 F NFT
to British Salt Ltd, Middlewich, Cheshire, May 1973; to Staffordshire Locomotives and stored at J. & H. Parry & Sons Ltd, Shawbury, Shropshire, 13th April 2000; to Cotswold Rail, c/o Allelys, Studley, Warwickshire, September 2000; to Fire Services College, Moreton in Marsh, about December 2000; to Cotswold Rail, RAF Quedgeley, near Gloucester, by 2nd July 2001; used for spares; remains to European Metal Recycling, Snailwell, Cambridgeshire, 2001; scrapped, July 2001.

D2152 Swindon 1960 87E 10/83 P D2152
03152 to Swindon & Cricklade Railway; despatched from BR Swindon Works, 6th March 1986; to Swindon Heritage Centre, April 1988; displayed at Membury Services West (on the M4) from about April 1990; returned to Swindon Heritage Centre, November 1990; to Swindon & Cricklade Railway, about March 1993.

D2158 Swindon 1960 NC 6/87 P D2158 / MARGARET ANN
03158
to Knights of Old Ltd, Old, Northamptonshire; despatched from 31B March Depot, 25th January 1992; to Peak Rail, Darley Dale, 3rd January 1997; to Ecclesbourne Valley Railway, Wirksworth, 7th September 2004; to Lincolnshire Wolds Railway, Ludborough, 16th June 2009; to Great Central Railway, Ruddington, Nottingham, 12th August 2009; to Titley Junction Station, near Kington, Herefordshire, 3rd July 2014.

D2162 Swindon 1960 BD 1989 P 03162
03162 purchased by Wirral Borough Council and moved for storage to BR Chester Depot, 30th March 1989; to Llangollen Railway; despatched from BR Chester Depot, October 1989.

D2170 Swindon 1960 BD 5/89 P 03170
03170 to BR Chester for storage, 30th March 1989; to Otis Euro Trans Rail Ltd, Manchester; despatched from BR Chester Depot, 24th July 1989; to Harry Needle Railroad Company, December 1999; to Fragonset, Derby, for certification, about April 2000; returned to Otis Euro Trans Rail Ltd; to Barrow Hill Engine Shed Society, Staveley, 28th September 2000; to Battlefield Line, Shackerstone, 13th August 2001; to Epping Ongar Railway, Essex, 17th September 2010.

D2176 Swindon 1961 ZC 5/68 F D2176
to George Cohen, Sons & Co Ltd, Cransley, Northamptonshire, October 1968; scrapped, November 1971.

D2178 Swindon 1962 81F 9/69 P D2178
to A.R. Adams & Son, Newport, Monmouthshire, January 1970; despatched from 81F Oxford Depot; used as a hire locomotive (see Appendix A); sold to National Smokeless

Fuels Ltd, Coed Ely Coking Plant, Tonyrefail, May 1974; to BR Swindon Works, for repairs, by July 1979; still at Swindon Works, 12th August 1979; seen at BR Severn Tunnel Junction Depot (on way back to Coed Ely Coking Plant, Tonyrefail), 1st November 1979; to Caerphilly Railway Preservation Society, Caerphilly, 12th November 1985; to Gwili Railway, Bronwydd Arms, 21st October 1996; to St Philip's Marsh Depot, Bristol, for tyre turning, 7th December 2011; returned to Gwili Railway, Bronwydd Arms, 14th December 2011.

D2179 Swindon 1962 ? ? P 03179 / CLIVE

03179 to West Anglia & Great Northern Company, Electric Maintenance Depot, Hornsey, 8th June 1998; to Nene Valley Railway, Wansford, for gala, 29th February 2008; returned to First Capital Connect, Hornsey, early March 2008; to Rushden Historical Transport Society, Rushden, Northamptonshire, 19th July 2016.

D2180 Swindon 1962 NC 3/84 P 03180

03180 to Mayer Newman Ltd, Snailwell, Cambridgeshire, 26th July 1984; to South Yorkshire Railway Preservation Society (HNRC) Meadowhall, Sheffield, South Yorkshire, 21st December 1991; to Battlefield Line, Shackerstone, 2nd August 2001; to Peak Rail, Rowsley, 14th November 2011.

D2181 Swindon 1962 87E 5/68 F PRIDE OF GWENT

to A.R. Adams & Son, Newport, Monmouthshire; despatched from 85A Worcester Depot, 10th December 1968; used as a hire locomotive (see Appendix A); sold to Gwent Coal Distribution Centre, Newport, Monmouthshire, by August 1971; to Marple & Gillott Ltd, Attercliffe, Sheffield, for scrap, December 1986; scrapped, January 1987.

D2182 Swindon 1962 87E 5/68 P D2182

to A.R. Adams & Son, Newport, Monmouthshire; despatched from 85A Worcester Depot, 29th November 1968; used as a hire locomotive (see Appendix A); sold to Lindley Plant Ltd, Gatewen Disposal Point, Denbighshire, September 1973; to NCBOE Bennerley Disposal Point, 1981; to NCBOE Wentworth Stores, Rotherham, 18th March 1982; to NCBOE Bennerley Disposal Point, Ilkeston, 6th May 1983; to NCBOE Coalfield Farm Disposal Point, Hugglescote, Leicestershire, July 1983; to Warwick District Council, Victoria Park, Leamington, 20th April 1986; to Gloucestershire Warwickshire Railway, Toddington, 11th January 1993.

D2184 Swindon 1962 87E 12/68 P D2184

to Co-operative Wholesale Society Ltd, Coal Concentration Depot, Southend-on-Sea, Essex; despatched from 85A Worcester Depot, 20th August 1969; to Colne Valley Railway, Castle Hedingham, Essex, 17th October 1986.

D2185 Swindon 1962 85A 12/68 F D2185

to Bird's Commercial Motors Ltd, Long Marston, Worcestershire; despatched from 85A Worcester Depot, 21st May 1969; seen at Long Marston, 23rd March 1970; to Abercarn Tinplate Works, April 1970; seen at Abercarn, 9th October 1971; to Bird's (Swansea) Ltd, 40 Acre Site, Cardiff, by 16th July 1972; to Bird's, Long Marston, January 1978; scrapped, June 1978.

D2186 Swindon 1962 81F 9/69 F D2186

to A.R. Adams & Son, Newport, Monmouthshire; despatched from 81F Oxford Depot, 6th February 1970; arrived at Adams, 8th February 1970; retained its BR livery and number whilst with Adams; used as a hire locomotive (see Appendix A); scrapped, January 1981.

D2187 Swindon 1961 82C 5/68 F NFT

to Bird's Commercial Motors Ltd, Long Marston, Worcestershire; despatched from BR Swindon Depot, September 1968; seen re-painted in yellow livery, with BR number painted over, 25th January 1970; scrapped, June 1978.

D2188 Swindon 1961 83B 5/68 F D2188

to Bird's Commercial Motors Ltd, Long Marston, Worcestershire; despatched from 82C Swindon Depot, September 1968; scrapped, February 1978.

D2189 Swindon 1961 6A 3/86 P 03189

03189 to Steamport, Southport (despatched from 31B March Depot), December 1991; to Ribble Steam Railway, Preston, 17th April 1999.

D2192 Swindon 1961 82C 1/69 P D2192 / TITAN

to Dart Valley Railway, Devon, 25th August 1970; to Torbay & Dartmouth Railway, Paignton, 24th July 1977; to South Devon Railway Trust, Buckfastleigh, summer 1991; to Paignton and Dartmouth Steam Railway, Devon, 25th July 1991.

D2193 Swindon 1961 82C 1/69 F 2

to A.R. Adams & Son, Newport, Monmouthshire; despatched from 85A Worcester Depot, September 1969; used as a hire locomotive (see Appendix A); scrapped, January 1981.

D2194 Swindon 1961 85A 9/68 F D2194

to Bird's Commercial Motors Ltd, Long Marston, Worcestershire; despatched from 85A Worcester Depot, about May 1969; scrapped, about August 1978.

D2195 Swindon 1961 82A 9/68 F D10

to Llanelly Steel Co Ltd, Carmarthenshire (sold per R.E. Trem Ltd, Finningley, Doncaster); despatched from 85A Worcester Depot, about May 1969; scrapped, September 1981.

D2196 Swindon 1961 8H 6/83 P 03196 / 40 / JOYCE

03196 to R.O. Hodgson Ltd, Carnforth, Lancashire, 15th June 1983; to Steamtown, Carnforth, by July 1992.

D2197 Swindon 1961 NC 6/87 P 03197

03197 to South Yorkshire Railway Preservation Society (HNRC), Meadowhall, Sheffield; despatched from 15A Leicester Depot, 25th October 1991; to Lavender Line, Isfield, for storage, 31st July 2001; to Sonic Rail Ltd, Burnham-on-Crouch, Essex, for overhaul, 13th December 2010; to Mangapps Farm Railway Museum, Burnham-on-Crouch, 24th August 2012; to Isle of Wight Railway, 25th September 2016; to Mid-Norfolk Railway, Dereham, 3rd October 2018.

D2199 Swindon 1961 12C 6/72 P D2199

to BR Doncaster Works, for overhaul and fitting with air brakes, summer 1973; seen at

Doncaster Works, 16th February 1974; to Rockingham Colliery, Birdwell, Barnsley, February 1974; to Barrow Colliery, Worsborough, Barnsley, about January 1979; to Houghton Main Colliery, Barnsley, about June 1979; to Royston Drift Mine, Barnsley, 14th August 1980; to Barrow Colliery, Worsborough, 8th July 1981; to Royston Drift Mine, 23rd March 1982; to Royston Machinery Stores, September 1986; to South Yorkshire Railway Preservation Society, Attercliffe, Sheffield, 12th August 1987; to SYRPS, Meadowhall, Sheffield, 12th September 1988; displayed at BR Tinsley Depot, Sheffield, open day, 29th September 1990; to RMS Locotec, for use at Euro Tunnel, Cheriton Terminal, on hire, 14th March 1997; returned to SYRPS, Meadowhall, Sheffield, 12th September 1997; to RMS Locotec, Dewsbury, on hire, about January 2001; returned to SYRPS, 2001; to Hanson Quarry Products, Machen, near Newport, on hire, 5th February 2001; to Peak Rail, Rowsley, 6th April 2006.

D2371 Swindon 1958 52A 12/87 P D2371

03371 to A.J. Wilkinson, Rowden Mill Station, near Bromyard, Worcestershire, 10th November 1988; to Dartmouth Steam Railway, Devon, 2nd February 2015.

D2373 Swindon 1961 9D 5/68 F No.1 / DAWN

to NCB Manvers Main Coal Preparation Plant, Wath upon Dearne, Rotherham; despatched from BR Bolton Depot, September 1968; scrapped on site by Ernest Nortcliffe & Son Ltd of Rotherham, by 13th July 1982.

D2381 Swindon 1961 16C 6/72 P D2381 / 03381

overhauled and re-painted at BREL Derby, early 1973; seen at Etches Park Depot, Derby, 9th April 1973; to Flying Scotsman Enterprises, Market Overton, Rutland, by rail, 13th April 1973; to Steamtown, Carnforth, 19th March 1976.

D2397 Doncaster 1961 NC 7/87 F 03397

03397 to The Vic Berry Company, Leicester; despatched from 31B March Depot, 23rd March 1990; used for spares; remains scrapped, January 1991.

D2399 Doncaster 1961 NC 7/87 P 03399

03399 to Mangapps Farm Railway Museum, Burnham-on-Crouch; despatched from 31B March Depot, 22nd March 1989; to Isle of Wight Railway, Havenstreet, 26th September 2018.

SECTION 2

Drewry Car Co Ltd 0-6-0 diesel mechanical locomotives built by Vulcan Foundry Ltd, numbered D2200-D2214, and introduced 1952. Fitted with a Gardner 8L3 engine developing 204bhp at 1200rpm, five speed gearbox, and driving wheels of 3ft 3in diameter. Later classified TOPS Class 04.

D2203 DC 2400 1952 ZC 12/67 P 11103
VF D145

to Hemel Hempstead Lightweight Concrete Co Ltd, Cupid Green, Hertfordshire, 2nd February 1968; to Embsay & Bolton Abbey Railway, 8th February 1982.

D2204 DC 2485 1953 55F 10/69 F D5
VF D211

to Briton Ferry Steel Co Ltd, Glamorgan, March 1970 (sold via W. & F. Smith Ltd, Ecclesfield, Sheffield); scrapped, September 1979.

D2205 DC 2486 1953 51L 7/69 P D2205
VF D212

to Tees & Hartlepool Port Authority, Middlesbrough Docks, July 1970; to T&HPA, Grangetown Docks, about September 1980; to Kent & East Sussex Railway, Tenterden, 21st August 1983; to West Somerset Railway, Minehead, 18th November 1989; to Somerset & Avon Railway, Radstock, 2nd February 1994; to West Somerset Railway, Minehead, July 1996; to Peak Rail, Rowsley, 14th October 2012.

D2207 DC 2482 1953 ZC 12/67 P D2207
VF D208

to Hemel Hempstead Lightweight Concrete Co Ltd, Cupid Green, Hertfordshire, February 1968; to North Yorkshire Moors Railway, Grosmont, September 1973; to RMS Locotec, Dewsbury, for overhaul, 22nd February 2005; to RMS Locotec, Wakefield, for further overhaul, 29th June 2006; to North Yorkshire Moors Railway, Pickering, 31st January 2007.

D2208 DC 2483 1953 5A 7/68 F D2208
VF D209

to NCB Manvers Coal Preparation Plant, Wath upon Dearne, Rotherham; despatched from 5B Crewe South Depot, 10th November 1968; to Cortonwood Colliery, Wombwell, March 1969; to Cadeby Colliery, Conisbrough, by May 1969; to Silverwood Colliery, Thrybergh, about December 1970; dismantled June 1976; scrapped by a dealer from Worksop, by January 1979.

D2209 DC 2484 1953 8J 7/68 F No.16 / TRACEY
VF D210

to NCB Manvers Coal Preparation Plant, Wath upon Dearne, Rotherham; despatched from 8J Allerton Depot, 10th November 1968; to Kiveton Park Colliery, 13th July 1974; used for spares, 1982; remains scrapped on site by Brinsworth Metals Ltd, 19th August 1985.

D2211 DC 2509 1954 16C 7/70 F WILF CLEMENT
VF D243

despatched from BR Derby Depot to Carmarthen Station; then to Powell Duffryn Fuels Ltd, NCBOE Coed Bach Disposal Point, Kidwelly, August 1970; to Rees Industries Ltd, Llanelli, 3rd August, 1978; scrapped, about November 1980.

D2213 DC 2529 1954 8H 8/68 F D2213
VF D257

to NCB Manvers Coal Preparation Plant, Wath upon Dearne, Rotherham, September 1969; used for spares, by July 1975; remains scrapped, February 1978.

SECTION 3

Drewry Car Co Ltd 0-6-0 diesel mechanical locomotives built by Vulcan Foundry Ltd and Robert Stephenson & Hawthorns Ltd, numbered D2215-D2273, and introduced 1955. Fitted with a Gardner 8L3 engine developing 204bhp at 1200rpm, five speed gearbox, and driving wheels of 3ft 6in diameter. Later classified TOPS Class 04.

D2219 DC 2542 1955 8H 4/68 F D2219
VF D268

to Barnsley District Coking Co Ltd, Barrow Coking Plant, Barnsley, October 1968; to NCB Barrow Colliery, Barnsley, on loan, August 1969; returned to Barrow Coking Plant; scrapped, May 1977.

D2225 DC 2548 1955 8F 3/69 F D2225 / DEBRA
VF D274

to NCB Manvers Coal Preparation Plant, Wath upon Dearne, Rotherham, January 1970; to Wath Colliery, 8th December 1976; scrapped on site by Wath Skip Hire Ltd of Rotherham, July 1985.

D2228 DC 2551 1955 8F 7/68 F D2228 / 4
VF D277

to Bowaters UK Paper Co Ltd, Sittingbourne, Kent, 17th February 1969; scrapped, January 1979.

D2229 DC 2552 1955 52A 12/69 P D2229
VF D278

to NCB Brookhouse Colliery, Beighton; despatched from 51L Thornaby Depot, 28th August 1970; to Orgreave Colliery, by 28th July 1971; to Brookhouse Colliery, 12th March 1972; to Orgreave Colliery about October 1973; to Brookhouse Colliery by July 1974; to Manton Colliery, 28th March 1983; to South Yorkshire Railway Preservation Society, Meadowhall, Sheffield, 26th May 1990; to Peak Rail, Rowsley, 12th March 2002.

D2238 DC 2562 1955 8H 7/68 F D2238 / CAROL
VF D288

to NCB Manvers Coal Preparation Plant, Wath upon Dearne, Rotherham, 10th November 1968; to Manvers Coking Plant, Wath upon Dearne, about 1971; to Coventry Home Fire Plant, Keresley, on loan (as cover while HE 6658 was at Hunslet Engine Company, Leeds, for repairs), 15th June 1974; returned to Manvers Coking Plant, Wath upon Dearne, 5th December 1975; ovens closed down, January 1981; scrapped on site by Ernest Nortcliffe & Son Ltd of Rotherham, by 13th July 1982.

D2239 DC 2563 1955 75C 9/71 F NFT
VF D289

to NCB Dodworth Colliery, Barnsley, September 1972; to C.F. Booth Ltd, Rotherham, for scrap, 20th March 1986; scrapped, March 1986.

D2241 DC 2565 1956 30E 5/71 F 2241
VF D291

to George Cohen, Sons & Co Ltd, Cransley, Northamptonshire, September 1971; scrapped, November 1976.

D2243 DC 2575 1956 51L 7/69 F MD2
RSHN 7862

to Tees & Hartlepool Port Authority, Middlesbrough Docks, July 1970; dismantled 1972; scrapped March 1973.

D2244 DC 2576 1956 55F 6/70 F 5
RSHN 7863

to A.R. Adams & Son, Newport, Monmouthshire, August 1970; despatched from BR Hammerton Street Depot, Bradford, but delayed en-route as ran hot at Belper; used by Adams as a hire locomotive (see Appendix A); scrapped, January 1981.

D2245 DC 2577 1956 50D 12/68 P D2245
RSHN 7864

to Derwent Valley Railway Company, Layerthorpe, York, May 1969; to Battlefield Line, Shackerstone, Leicestershire, 17th May 1978; to Derwent Valley Light Railway, Murton, York, 17th July 2013; to Battlefield Line, Shackerstone, Leicestershire, 22nd July 2013; to Derwent Valley Light Railway Society, Murton, York, 30th May 2014.

D2246 DC 2578 1956 55G 7/68 P D2246
RSHN 7865

to Coal Mechanisation Ltd, Crawley Coal Concentration Depot, Sussex; despatched from BR Knottingley Depot, January 1969; seen in a freight train en-route to Crawley, 17th January 1969; to Coal Mechanisation Ltd, Tolworth Coal Depot, Surrey, by 12th August 1982; to British Coal, West Drayton Landsale Depot, 19th November 1990; to South Yorkshire Railway Preservation Society (HNRC), Meadowhall, Sheffield, 19th December 1994; to Elsecar Steam Railway, near Barnsley, on hire, 20th April 1995; returned to SYRPS, 21st August 1996; to South Devon Railway, Buckfastleigh, 9th January 2001.

D2247 DC 2579 1956 55B 11/69 F D12
RSHN 7866

to Ford Motor Co Ltd, Dagenham, for short period, date unknown; returned to British Railways; to Briton Ferry Steel Co Ltd, Glamorgan, June 1970 (sold via W. & F. Smith Ltd, Ecclesfield, Sheffield); scrapped, September 1979.

D2248 DC 2580 1957 55F 6/70 F 2243 / No.18 / SUE
RSHN 7867

to NCB Manvers Coal Preparation Plant, Wath upon Dearne, Rotherham, June 1970; to Maltby Colliery, Rotherham, about September 1971; during a repaint at Maltby Colliery received the incorrect number 2243; scrapped by Carol & Good Ltd, Thurcroft, near Rotherham, April 1987.

D2258 DC 2602 1957 16C 9/70 F D2258 / 4-2
RSHD 7879

to Hargreaves Industrial Services Ltd, NCBOE Bennerley Disposal Point, Ilkeston, January 1971; to BR Toton Depot, for repairs, December 1974; returned to NCBOE Bennerley Disposal Point, Ilkeston, January 1975; to NCBOE Wentworth Stores, near Rotherham, 17th February 1984; to C.F. Booth Ltd, Rotherham, for scrap, 2nd September 1986; scrapped, January 1987.

D2259 DC 2603 1957 73F 12/68 F D2259 / 5
RSHD 7889

to Bowaters UK Paper Co Ltd, Sittingbourne, Kent, February 1969; seen outside the shed at Ridham Dock, being used for spares, 30th October 1976; remains scrapped on site by Smeeth Metal Co Ltd, January 1978.

D2260 DC 2604 1957 55F 10/70 F THOMAS HARLING
RSHD 7890

to Tilsley & Lovatt Ltd, Trentham, 11th March 1971; to Powell Duffryn Fuels Ltd, NCBOE Mill Pit Disposal Point, Cefn Cribbwr, July 1971; to Cwm Mawr Disposal Point, Tumble, 3rd November 1981; to Coed Bach Disposal Point, Kidwelly, December 1981; scrapped on site by Rees Industries Ltd of Llanelli, June 1983.

D2262 DC 2606 1957 51A 9/68 F 7
RSHD 7892

to Ford Motor Co Ltd, Dagenham, March 1969; involved in a collision; later used for spares; remains scrapped, July 1978.

D2267 DC 2611 1957 50D 12/69 F No.1
RSHD 7897

to Ford Motor Co Ltd, Dagenham, January 1970; to BR Swindon Works, for rebuild, 19th May 1977; to Ford Motor Co Ltd, Dagenham, 8th November 1977; dismantled (no engine) by October 1996; to East Anglian Railway Museum, Wakes Colne, Essex, 24th September 1998; to North Norfolk Railway, Sheringham, 16th February 2000; used for spares; remains scrapped, April 2003.

D2270 DC 2614 1957 55B 2/68 F D9
RSHD 7912

to Briton Ferry Steel Co Ltd, Glamorgan, July 1968 (sold via R.E. Trem Ltd, Finningley, Doncaster); seen freshly re-painted and lost its BR number, 7th July 1969; scrapped, September 1979.

D2271 DC 2615 1958 55F 10/69 P D2271
RSHD 7913

to C.F. Booth Ltd, Rotherham, May 1970; privately purchased for preservation and moved to Thomas Hill Ltd, Kilnhurst, for storage, 27th July 1972; to Midland Railway, Normanton Barracks, Derby, 7th September 1973; to Midland Railway, Butterley, 10th May 1975; to West Somerset Railway, Minehead, 15th May 1982; to South Devon Railway, Buckfastleigh, 1st November 2018.

D2272 DC 2616 1958 55F 10/70 P D2272 / ALFIE
RSHD 7914

to British Fuel Company, Coal Concentration Depot, Blackburn, March 1971; to South Yorkshire Railway Preservation Society (HNRC), Meadowhall, Sheffield, 1st May 1997; to Lavender Line, Isfield, about June 2001; to Peak Rail, Rowsley, February 2004.

SECTION 4

Drewry Car Co Ltd 0-6-0 diesel mechanical locomotives built by Robert Stephenson & Hawthorns Ltd, numbered D2274-D2340, and introduced 1959. Fitted with a Gardner 8L3 engine developing 204bhp at 1200rpm, five speed gearbox, and driving wheels of 3ft 7in diameter. Later classified TOPS Class 04. Departmental DS1173, built in 1947, was later numbered D2341 and completed Class 04, but some particulars were different from those of the main batch.

D2274 DC 2620 1959 8J 5/69 F D2274 / No.17
RSHD 7918

to NCB Maltby Colliery, Rotherham; despatched from 8J Allerton Depot, 24th June 1969; scrapped, September 1980.

D2276 DC 2622 1959 30A 8/69 F D2276
RSHD 7920

to A.R. Adams & Son, Newport, Monmouthshire, July 1970; used for spares; remains scrapped, May 1977.

D2279 DC 2656 1960 30E 5/71 P D2279
RSHD 8097

to CEGB Rye House Power Station, Hoddesdon, Hertfordshire, 1st October 1971; to East Anglian Railway Museum, Wakes Colne, Essex, about March 1981; to Peak Rail, Rowsley, for repairs, 31st January 2014; to Andrew Briddon, Darley Dale, 29th May 2015; to haulage yard, near Ashbourne, for storage, 14th June 2018; to East Anglian Railway Museum, Wakes Colne, Essex, 4th July 2018.

D2280 DC 2657 1960 30E 3/71 P D2280
RSHD 8098

to Ford Motor Co Ltd, Dagenham; despatched from 30E Colchester Depot, 28th June 1971; to BR Swindon Works, for rebuild, 8th July 1977; to Ford Motor Co Ltd, Dagenham, 8th November 1977; to East Anglian Railway Museum, Wakes Colne, Essex, 24th September 1998; to North Norfolk Railway, Sheringham, 16th February 2000; to Gloucestershire Warwickshire Railway, Toddington, 22nd March 2018.

D2281 DC 2658 1960 30E 10/68 F D2281
RSHD 8099

to Briton Ferry Steel Co Ltd, Glamorgan, February 1969 (sold via R.E. Trem Ltd, Finningley, Doncaster); used for spares; remains scrapped, August 1971.

D2284 DC 2661 1960 30E 4/71 P D2284
RSHD 8102

to NCB North Gawber Colliery, Mapplewell, Barnsley, 16th July 1971; to Grimethorpe Colliery, 30th January 1976; to Woolley Colliery, Darton, early March 1978; to South Yorkshire Railway Preservation Society, Chapeltown, 2nd August 1985; to SYRPS, Attercliffe, Sheffield, December 1986; to SYRPS, Meadowhall, Sheffield, 12th September 1988; to Peak Rail, Rowsley, March 2002.

D2289 DC 2669 1960 70D 9/71 P 3945
RSHD 8122

to Peak Rail, Rowsley, 13th June 2018 (from Appendix C).

D2294 DC 2674 1960 70D 2/71 F 01
RSHD 8127

to Shipbreaking (Queenborough) Ltd, Kent, 16th March 1972; seen re-painted in blue livery and lost its BR number, working at quay, 22nd October 1972; scrapped, October 1985.

D2298 DC 2679 1960 52A 12/68 P D2298
RSHD 8157

to Derwent Valley Railway Company, Layerthorpe, York, April 1969; to Quainton Railway Society, near Aylesbury, Buckinghamshire, 22nd October 1982; to East Lancashire Railway, Bury, for gala, 25th May 2017; returned to Quainton Railway Society, 7th June 2017.

D2299 DC 2680 1960 52A 1/70 F D2299 / JONAH
RSHD 8158

to NCB Bestwood Colliery, Nottinghamshire; despatched from BR Thornaby Depot, 9th July 1970; to Hucknall Colliery, 7th August 1970; to Calverton Colliery, 14th November 1977; to Hucknall Colliery, 23rd August 1978; to C.F. Booth Ltd, Rotherham, for scrap, February 1984; scrapped, week-ending 17th February 1984.

D2300 DC 2681 1960 8J 5/69 F D2300
RSHD 8159

to NCB Shireoaks Colliery, 25th June 1969; to Steetley Colliery, on loan, 12th September 1974; returned to Shireoaks Colliery, 18th November 1974; to Manton Colliery, 18th October 1978; scrapped by Hoyland Dismantling Co Ltd, August 1986.

D2302 DC 2683 1960 16C 6/69 P D2302
RSHD 8161

to British Sugar Corporation Ltd, Woodston Factory, Peterborough, August 1969; to BSC Allscott Factory, Shropshire, October 1969; to G.G. Papworth Ltd, Queen Adelaide Rail Distribution Centre, Ely, Cambridgeshire, 12th July 1983; site taken over by Potter Group, 1991; to South Yorkshire Railway Preservation Society (HNRC), Meadowhall, Sheffield, 25th September 1993; to Rutland Railway Museum, Cottesmore, on loan, 16th March 2001; to Barrow Hill Engine Shed Society, Staveley, 18th May 2004; to D2578 Locomotive Group, Moreton on Lugg, 16th November 2011.

D2304 DC 2685 1960 51A 2/68 F D8
RSHD 8163

to Llanelly Steel Co Ltd, Carmarthenshire (sold via R.E. Trem Ltd, Finningley, Doncaster), July 1968; scrapped, May 1977.

D2305 DC 2686 1960 51A 2/68 F D9
RSHD 8164

to Llanelly Steel Co Ltd, Carmarthenshire (sold via R.E. Trem Ltd, Finningley, Doncaster), May 1968; scrapped, September 1981.

D2306 DC 2687 1960 51L 2/68 F D6
RSHD 8165

to Llanelly Steel Co Ltd, Carmarthenshire (sold via R.E. Trem Ltd, Finningley, Doncaster), 11th July 1968; scrapped, September 1981.

D2307 DC 2688 1960 51L 2/68 F D7
RSHD 8166

to Llanelly Steel Co Ltd, Carmarthenshire (sold via R.E. Trem Ltd, Finningley, Doncaster), July 1968; scrapped, October 1979.

D2308 DC 2689 1960 51A 2/68 F D8
RSHD 8167

to Briton Ferry Steel Co Ltd, Glamorgan (sold via R.E. Trem Ltd, Finningley, Doncaster), August 1968; seen passing through Rotherham on a low-loader, 24th August 1968; seen still carrying its BR number, 7th July 1969 and 14th September 1971; to Duport Steel Works Ltd, Llanelli, 25th October 1979; scrapped, May 1980.

D2310 DC 2691 1960 52A 1/69 P 04110
RSHD 8169

to Coal Mechanisation Ltd, Tolworth Coal Depot, Surrey, May 1969; to South Yorkshire Railway Preservation Society (HNRC), Meadowhall, Sheffield, 14th September 1994; to Battlefield Line, Shackerstone, 3rd October 2001; sold to Battlefield Line, about October 2011.

D2317 DC 2698 1960 52A 8/69 F No.10
RSHD 8176

to NCB Manvers Main Coal Preparation Plant, Wath upon Dearne, Rotherham, 30th December 1969; to Cortonwood Colliery, Wombwell, about 5th May 1970; scrapped on site by Wath Skip Hire Ltd of Rotherham, July 1986.

D2322 DC 2703 1961 52A 8/68 F D2322 / No.24
RSHD 8181

to NCB Orgreave Colliery, February 1969; thereafter used mainly at Orgreave Colliery although it sometimes worked through to Treeton Colliery along a private NCB branch line; to New Stubbin Colliery, Rawmarsh, on loan for about one month, summer 1975; returned to Orgreave Colliery; to Kiveton Park Colliery, 29th April 1980; scrapped by 28th November 1985.

D2324 DC 2705 1961 55B 7/68 P 2324 / JUDITH
RSHD 8183

to G.W. Talbot Ltd, Coal Concentration Depot, Aylesbury, Buckinghamshire, January 1969; to Redland Roadstone Ltd, Barrow upon Soar, Leicestershire, after 18th November 1989; to South Yorkshire Railway Preservation Society (HNRC), Meadowhall, Sheffield, 29th March 1995; to Lavender Line, Isfield, about June 2001; to Barrow Hill Engine Shed Society, Staveley, 27th March 2006; to Peak Rail, Rowsley, 1st October 2008; sold by HNRC, about October 2011; to Nemesis Rail, Burton upon Trent, early January 2015.

D2325 DC 2706 1961 50D 7/68 P D2325
RSHD 8184

to NCB Norwich Coal Concentration Depot, December 1968; to Tannick Commercial Repairs, Norwich, for storage, November 1986; to John Jolly, Bridgewick Farm, Dengie, Southminster, Essex, 19th March 1987; to Mangapps Farm Railway Museum, Burnham-on-Crouch, 19th March 1989.

D2326 DC 2707 1961 52A 8/68 F D2326
RSHD 8185

to NCB Manvers Main Coal Preparation Plant, Wath upon Dearne, Rotherham; despatched from 52A Gateshead Depot, February 1969; used for spares in 1971; remains scrapped on site, Autumn 1975.

D2327 DC 2708 1961 52A 8/68 F No.12 / 521-12
RSHD 8186

to NCB Manton Main Colliery, 1969; to Dinnington Colliery, 9th August 1971; to Elsecar Central Workshops, 3rd May 1973; returned to Dinnington Colliery, 1973; to Elsecar Central Workshops, 15th November 1974; to Dinnington Colliery, 20th January 1975; to Coopers (Metals) Ltd, Brightside, Sheffield, for scrap, 5th January 1984; scrapped, February 1984.

D2328 DC 2709 1961 52A 9/68 F No.31
RSHD 8187

to NCB Dinnington Colliery; despatched from 52A Gateshead Depot, 6th June 1969; to Steetley Colliery, April 1973; to BR Doncaster Works, for tyre turning, 2nd February 1977; returned to Steetley Colliery, 10th February 1977; to Shireoaks Colliery, by April 1982; to Kiveton Park Colliery, 13th May 1982; to Cortonwood Colliery, Wombwell, 18th July 1985; scrapped on site by Wath Skip Hire Ltd of Rotherham, July 1986.

D2329 DC 2710 1961 52A 7/68 F D2329
RSHD 8188

to Derwent Valley Railway Company, Layerthorpe, York, January 1969 (sold via Peter Wood & Co Ltd, Eckington, Sheffield); used for spares; remains scrapped, April 1970.

D2332 DC 2713 1961 52A 6/69 F D2332 / LLOYD
RSHD 8191

to NCB Manvers Main Coal Preparation Plant, Wath upon Dearne, Rotherham, January 1970; to Cadeby Colliery, Conisbrough, 28th August 1975; to Thurcroft Colliery, 14th June

1976; to Shireoaks Colliery, 29th June 1981; to Thurcroft Colliery, 3rd September 1982; to Dinnington Colliery, 19th July 1985; scrapped, July 1986.

D2333 DC 2714 1961 52A 9/69 F 3 / P1062C
RSHD 8192

to Ford Motor Co Ltd, Dagenham, 2nd March 1970; to BR Swindon Works, for rebuild, 3rd May 1977; to Ford Motor Co Ltd, Dagenham, 23rd January 1978; scrapped, early 1990.

D2334 DC 2715 1961 51A 7/68 P D2334
RSHD 8193

to NCB Manvers Main Coal Preparation Plant, Wath upon Dearne, Rotherham; despatched from 51A Darlington Depot, 2nd June 1969; to Thurcroft Colliery, 8th October 1969; to Dinnington Colliery, 19th July 1985; to Maltby Colliery, Rotherham, 24th February 1986; to South Yorkshire Railway Preservation Society, Meadowhall, Sheffield, 12th November 1988; to Knights of Old Ltd, Old, Northamptonshire, 28th September 1993; to Churnet Valley Railway, Cheddleton, Staffordshire, 10th July 1994; to Mid-Norfolk Railway, Dereham, 7th January 2017.

D2335 DC 2716 1961 51A 7/68 F No.2
RSHD 8194

to NCB Manvers Main Coal Preparation Plant, Wath upon Dearne, Rotherham; despatched from 51A Darlington Depot, 2nd June 1969; to Maltby Colliery, Rotherham, about September 1969; scrapped, September 1980.

D2336 DC 2717 1961 51A 7/68 F D2336
RSHD 8195

to NCB Manvers Main Coal Prepaation Plant, Wath upon Dearne, Rotherham; despatched from 51A Darlington Depot, 2nd June 1969; used for spares; remains scrapped, about February 1978.

D2337 DC 2718 1961 51A 7/68 P D2337
RSHD 8196

to NCB Manvers Main Coal Preparation Plant, Wath upon Dearne, Rotherham; despatched from 51A Darlington Depot, 2nd June 1969; to Barnburgh Main Colliery, June 1974; to Manvers Main Coal Preparation Plant, February 1977; to South Yorkshire Railway Preservation Society, Attercliffe, Sheffield, 22nd February 1988; to SYRPS, Meadowhall, Sheffield, 15th September 1988; to Peak Rail, Rowsley, 15th March 2002.

D2340 DC 2593 1956 55F 10/68 F D1
RSHD 7870

Demonstration locomotive; sold to British Railways, March 1962; to Briton Ferry Steel Co Ltd, Glamorgan, April 1969; seen still carrying its BR number, 7th July 1969; scrapped, September 1979.

SECTION 5

Andrew Barclay, Sons & Co Ltd built 0-4-0 diesel mechanical locomotives, numbered D2410-D2444, and introduced 1958. Fitted with a Gardner 8L3 engine developing 204bhp at 1200rpm; five speed gearbox, and driving wheels of 3ft 7in diameter. Later classified TOPS Class 06.

D2420 AB 435 1959 RSD 1984 P NPT
06003 withdrawn as BR Departmental locomotive 97804 at Reading Signal Depot; to C.F. Booth Ltd, Rotherham; despatched from 81A Old Oak Common Depot, 25th September 1986; to South Yorkshire Railway Preservation Society (HNRC), Attercliffe, Sheffield, 9th March 1987; to SYRPS, Meadowhall, Sheffield, 14th September 1988; displayed at BR Tinsley Depot open day, 29th September 1990; returned to SYRPS; to Crewe Works open day, 2nd May 1997; to Battlefield Line, Shackerstone, on loan, 9th May 1997; returned to SYRPS, 23rd October 1997; to Barrow Hill Engine Shed Society, Staveley, 11th June 1999; to Rutland Railway Museum, Cottesmore, 9th September 1999; to Barrow Hill Engine Shed Society, Staveley, 13th September 2002; to UK Coal, Widdrington Disposal Point, Northumberland, on hire, 18th July 2003; to Barrow Hill Engine Shed Society, Staveley, 28th June 2006; to Peak Rail, Rowsley, May 2008; to Museum of Science & Industry, Manchester, on loan, 14th November 2011; to Peak Rail, Rowsley, 31st January 2013.

D2432 AB 459 1960 65A 12/68 F NFT
to Shipbreaking (Queenborough) Ltd, Kent, 20th May 1969; seen working, 11th July 1972; exported from Sheerness Docks to Italy, March 1977. (see Appendix C).

SECTION 6

Hudswell, Clarke & Co Ltd built 0-6-0 diesel mechanical locomotives, numbered D2510-D2519, and introduced 1961. Fitted with a Gardner 8L3 engine developing 204bhp at 1200rpm; four speed gearbox, and driving wheels of 3ft 6in diameter. No TOPS classification.

D2511 HC D1202 1961 12C 12/67 P D2511 / BRM5477
to NCB Brodsworth Colliery, Doncaster; despatched from 12C Barrow Depot, May 1968; to Keighley & Worth Valley Railway, Haworth, 8th October 1977.

D2513 HC D1204 1961 12C 8/67 F D2513
to NCB Cadeby Colliery, Conisbrough; despatched from 12C Barrow Depot, December 1968; scrapped, October 1975.

D2518 HC D1209 1962 5A 2/67 F D2518 / 3219-016
to NCB Hatfield Colliery, Stainforth, Doncaster; despatched from 5A Crewe Depot, August 1967; later used for spares; written off, 3rd May 1974; remains scrapped on site by NCB, by April 1976.

D2519 HC D1210 1962 5A 7/67 F D2519 / 3219-017
to NCB Hatfield Colliery, Stainforth, Doncaster; despatched from 5A Crewe Depot,

February 1968; to Keighley & Worth Valley Railway, Haworth, 3rd April 1982; to Marple & Gillott Ltd, Attercliffe, Sheffield, for scrap, 27th March 1985; scrapped, April 1985.

SECTION 7

Hunslet Engine Co Ltd built 0-6-0 diesel mechanical locomotives, numbered D2550-D2618, and introduced 1955. Fitted with a Gardner 8L3 engine developing 204bhp at 1200rpm, four speed gearbox, and driving wheels of 3ft 4in diameter (D2550-D2573) and 3ft 9in (D2574-D2618). Later classified TOPS Class 05.

D2554 HE 4870 1956 70H 9/83 P D2554
05001 rebuilt with cut-down cab and worked as Departmental locomotive number 97803 on Isle of Wight; withdrawn by British Railways, September 1983; to Isle of Wight Steam Railway, Havenstreet, 27th August 1984.

D2561 HE 4999 1957 8F 8/67 F D3
to Llanelly Steel Co Ltd, Carmarthenshire, March 1968; scrapped, October 1972.

D2568 HE 5006 1957 8F 8/67 F D2568
to Briton Ferry Steel Co Ltd, Glamorgan, 9th May 1968; scrapped, about May 1969.

D2569 HE 5007 1957 8F 8/67 F D6
to Briton Ferry Steel Co Ltd, Glamorgan; despatched from 8C Speke Junction Depot, May 1968; scrapped, about May 1969.

D2570 HE 5008 1957 8F 7/67 F NFT
to Briton Ferry Steel Co Ltd, Glamorgan; despatched from 8C Speke Junction Depot, March 1968; used for spares; seen in dismantled state with no number, 7th July 1969; remains scrapped, June 1971.

D2578 HE 5460 1958 62A 7/67 P D2578
to Hunslet Engine Co Ltd, Leeds, December 1967; rebuilt as Hunslet 6999; to H.P. Bulmer Ltd, Cider Manufacturers, Hereford, July 1968; whilst at Bulmer's was named CIDER QUEEN; to private site, Croes Newydd, Wrexham, November 2000; purchased by D2578 Locomotive Group, May 2001; to D2578 Locomotive Group, Moreton on Lugg, 6th August 2001.

D2587 HE 5636 1959 62C 12/67 P D2587
to Hunslet Engine Co Ltd, Leeds, September 1968; rebuilt (Hunslet 7180 of 1969) with a 384hp engine; to CEGB Chadderton Power Station, September 1969; to CEGB Kearsley Power Station, 3rd November 1981; to East Lancashire Railway, Bury, March 1983; to South Yorkshire Railway Preservation Society (HNRC), Meadowhall, Sheffield, 30th August 1997; to Lavender Line, Isfield, about June 2001; to Peak Rail, Rowsley, February 2004; sold by Harry Needle Railroad Company, 2011; to Barrow Hill Engine Shed Society, Staveley, 14th November 2011; to Peak Rail, Rowsley, 15th February 2013.

D2595 HE 5644 1960 62A 6/68 P D2595
to Hunslet Engine Co Ltd, Leeds, January 1969; rebuilt (Hunslet 7179 of 1969) with a

384hp engine; to CEGB Chadderton Power Station, September 1969; to CEGB Kearsley Power Station, 24th September 1981; to East Lancashire Railway, Bury, March 1983; to Steamport Transport Museum, Southport, Merseyside, 19th October 1989; to Ribble Steam Railway, Preston, 31st March 1999.

D2598 HE 5647 1960 50D 12/67 F SAM
to NCB Rossington Colliery, Doncaster; despatched from 50D Goole Depot, 17th May 1968; to Askern Colliery, Doncaster, July 1971; suffered fire damage, 1974; to Lambton Engine Works, Tyne & Wear, February 1975; scrapped, May 1975.

D2599 HE 5648 1960 50D 12/67 F 3219-013
to NCB Hickleton Colliery, Doncaster; despatched from 50D Goole Depot, May 1968; to Frickley Colliery, South Elmsall, about October 1968; to Askern Colliery, Doncaster, 16th June 1976; scrapped on site by R.D. Geeson Ltd of Ripley, May 1981.

D2600 HE 5649 1960 50D 12/67 F D7
to Briton Ferry Steel Co Ltd, Glamorgan (sold via R.E. Trem Ltd, Finningley, Doncaster); despatched from 50D Goole Depot, 29th April 1968; seen freshly re-painted and lost its BR number, 7th July 1969; withdrawn at end of 1970; scrapped, June 1971.

D2601 HE 5650 1960 50D 12/67 F D5
to Llanelly Steel Co Ltd, Carmarthenshire (sold via R.E. Trem Ltd, Finningley, Doncaster); despatched from 50D Goole Depot, 29th April 1968; arrived at Llanelli, 3rd May 1968; scrapped, September 1979.

D2607 HE 5656 1960 6G 12/67 F D2607
to NCB Dinnington Colliery, 9th September 1968; to Steetley Colliery about October 1968; to Shireoaks Colliery, on loan, May 1971; returned to Steetley Colliery, August 1971; to NCB Fence Workshops, Woodhouse Mill, Sheffield, for overhaul, 28th May 1974; returned to Steetley Colliery, 30th October 1974; to Shireoaks Colliery, on loan, 17th May 1975; returned to Steetley Colliery, July 1975; to Treeton Colliery, on loan, 3rd December 1975; returned to Steetley Colliery, December 1975; to BR Doncaster Depot, for tyre turning, August 1977; returned to Steetley Colliery about September 1977; to Shireoaks Colliery, on loan, 26th August 1980; returned to Steetley Colliery, 16th January 1981; to Treeton Colliery, on loan, April 1981; returned to Steetley Colliery, July 1981; to Coopers (Metals) Ltd, Brightside, Sheffield, for scrap, 12th June 1984; scrapped, by 4th July 1984.

D2611 HE 5660 1960 50D 12/67 F D2611 / YM1835
to NCB Yorkshire Main Colliery, Edlington, Doncaster; despatched from 50D Goole Depot, May 1968; written off, 3rd August 1976; scrapped on site, about December 1976.

D2613 HE 5662 1960 50D 12/67 F D2613 / BRM5481
to NCB Brodsworth Colliery, Woodlands, Doncaster; despatched from 50D Goole Depot, May 1968; to Bentley Colliery, 1974; written off, 6th April 1977; scrapped on site by Walter Heselwood Ltd of Sheffield, June 1977.

D2616 HE 5665 1961 50D 12/67 F D2616
to NCB Hatfield Colliery, Stainforth, Doncaster; despatched from 50D Goole Depot, May 1968; dismantled in April 1973; scrapped by December 1973.

D2617 HE 5666 1961 62C 12/67 F D2617
to Hunslet Engine Co Ltd, Leeds, autumn 1968; used for spares; remains scrapped, April 1976.

SECTION 8

North British Locomotive Co Ltd built 0-4-0 diesel hydraulic locomotives, numbered D2708-D2780, and introduced 1957. Fitted with a North British/M.A.N. W6V 17.5/22A engine developing 225bhp at 1100rpm, and driving wheels of 3ft 6in diameter. No TOPS classification.

D2720 NB 27815 1958 64H 8/67 F NFT
to James N. Connell Ltd, Coatbridge, Lanarkshire, June 1968; scrapped, July 1971.

D2726 NB 27821 1958 ZN 2/67 F NFT
to Shipbreaking (Queenborough) Ltd, Kent (sold via R.E. Trem Ltd, Finningley, Doncaster), October 1967; withdrawn from service, about September 1971; engine removed and sold to R.E. Trem Ltd, Finningley, Doncaster; remains scrapped on site, October 1971.

D2736 NB 27831 1958 65A 3/67 F D2736
to Bird's Commercial Motors Ltd, Long Marston, Worcestershire; despatched from 65A Eastfield Depot, July 1967; to Bird's (Swansea) Ltd, Pontymister Works, Risca, August 1967; to Bird's (Swansea) Ltd, 40 Acre Site, Cardiff, 25th February 1968; seen on 24th October 1968 and 6th October 1969; scrapped shortly after.

D2738 NB 27833 1958 65A 6/67 F NFT
to Andrew Barclay, Sons & Co Ltd, Kilmarnock, October 1967; rebuilt 1968; to NCB Killoch Colliery, Ochiltree, about July 1969; sold to Alex Smith Metals of Ayr, by March 1979; scrapped on site, January 1980.

D2739 NB 27834 1958 65A 3/67 F D2739
to Bird's Commercial Motors Ltd, Long Marston, Worcestershire; despatched from 65A Eastfield Depot, July 1967; scrapped, September 1969.

D2757 NB 28010 1960 65A 7/67 F NFT
to Bird's (Swansea) Ltd, Pontymister Works, Risca, 9th November 1967; to Bird's (Swansea) Ltd, 40 Acre Site, Cardiff, February 1968; seen on 17th February 1969 and 6th October 1969; scrapped, October 1970.

D2763 NB 28016 1960 65A 6/67 F NFT
to Andrew Barclay, Sons & Co Ltd, Kilmarnock, October 1967; rebuilt 1968; to BSC Landore Foundry, Swansea, by September 1969; scrapped, April 1977.

D2767 NB 28020 1960 65A 6/67 P D2767
to Andrew Barclay, Sons & Co Ltd, Kilmarnock, October 1967; rebuilt 1968; to Burmah Oil Co Ltd, Stanlow, Cheshire, 24th April 1969; to East Lancashire Railway, Bury, 12th June 1983; to Manchester Metrolink, on hire, October 1991; returned to East Lancashire

Railway, Bury, late 1991; to Scottish Railway Preservation Society, Bo'ness, 25th July 2001.

D2774 NB 28027 1960 65A 6/67 P D2774
to Andrew Barclay, Sons & Co Ltd, Kilmarnock, October 1967; rebuilt 1968; to NCB Killoch Colliery, Ochiltree, on hire, about July 1969; returned to Andrew Barclay, Sons & Co Ltd, 1971; sold to NCB; to Celynen North Colliery, Newbridge, early April 1971; to BR Canton Depot, Cardiff, for repairs, March 1976; seen at Hafodyrynys Colliery, Pontypool, on 26th March 1976; to Celynen North Colliery, Newbridge, April 1976; to Celynen South Colliery, Abercarn, 4th September 1976; to BR Canton Depot, Cardiff, for tyre turning, 17th May 1982; to NCB Mountain Ash Works, 27th May 1982; to Celynen South Colliery, Abercarn, by 12th March 1983; to East Lancashire Railway, Bury, 4th October 1986; to Strathspey Railway, Aviemore, 2nd May 2001.

D2777 NB 28030 1960 65A 3/67 F D2777
to Bird's Commercial Motors Ltd, Long Marston, Worcestershire, July 1967; to Bird's (Swansea) Ltd, Pontymister Works, Risca, about 9th November 1967; used for spares; remains scrapped, May 1968.

SECTION 9

Yorkshire Engine Co Ltd built 0-4-0 diesel hydraulic locomotives, numbered D2850-D2869, and introduced 1960. Fitted with a Rolls-Royce C6NFL engine developing 179bhp at 1800rpm, and driving wheels of 3ft 6in diameter. Later classified TOPS Class 02.

D2853 YE 2812 1960 8J 6/75 P D2853 / PETER
02003 to L.C.P. Fuels Ltd, Shut End Works, West Midlands, 19th November 1975; displayed at BR Bescot Depot, Walsall, open day, 9th October 1988; returned to L.C.P. Fuels Ltd, October 1988; to South Yorkshire Railway Preservation Society (HNRC), Meadowhall, Sheffield, 18th April 1997; to Rutland Railway Museum, Cottesmore, 22nd June 2001; to Barrow Hill Engine Shed Society, Staveley, 1st September 2003; to Appleby Frodingham Railway Society, Scunthorpe, on loan, 24th March 2006; to Barrow Hill Engine Shed Society, Staveley, 28th January 2016.

D2854 YE 2813 1960 8J 2/70 P D2854
condemned by BR as 'surplus – wear and tear', week-ending 28th February 1970; to C.F. Booth Ltd, Rotherham, 29th August 1970; used as a yard shunter; to South Yorkshire Railway Preservation Society (HNRC), Attercliffe, Sheffield, May 1988; to SYRPS, Meadowhall, Sheffield, 14th September 1988; to Middleton Railway, Leeds, on loan, 10th September 1994; returned to SYRPS, 15th October 1994; to Supertram, Nunnery Depot, Sheffield, on hire, 7th November 1994; returned to SYRPS, 27th May 1995; to Peak Rail, Darley Dale, March 2002.

D2856 YE 2815 1960 8J 6/75 F 02004
02004 to Redland Roadstone Ltd, Mountsorrel, Leicestershire, 25th September 1975; used for spares, including removal of engine; remains to Budden Wood Quarry, Leicestershire, after 10th March 1977; scrapped on site by The Vic Berry Company of Leicester, 31st October 1978.

D2857 YE 2816 1960 8J 4/71 F NFT
to Bird's Commercial Motors Ltd, Long Marston, Worcestershire; despatched from 8J Allerton Depot, November 1971; seen on 31st March 1992; scrapped, June 1992.

D2858 YE 2817 1960 9A 2/70 P D2858
condemned by BR as 'surplus to requirements', week-ending 28th February 1970; to Hutchinson Estate & Dock Co (Widnes) Ltd, Widnes; despatched from 9D Newton Heath Depot, about 4th August 1970; to Fisons Fertilisers, Widnes, September 1978; to Lowton Metals Ltd, Haydock, 5th March 1981; to Butterley Engineering Ltd, Ripley, Derbyshire, November 1986; to Midland Railway, Butterley, Derbyshire, 14th June 2002.

D2860 YE 2843 1961 8J 12/70 P D2860
to Curator of Historical Relics, BR Preston Park, Brighton, March 1973; to National Railway Museum, York, 26th November 1977; to Thomas Hill (Rotherham) Ltd, Kilnhurst, for overhaul and repaint, 14th September 1978; to National Railway Museum, York, 3rd January 1979; to Gloucestershire Warwickshire Railway, Toddington, for gala, 9th July 2009; to National Railway Museum, York, mid-July 2009.

D2862 YE 2845 1961 10D 12/69 F ND3 / 63000359
to Tilsley & Lovatt Ltd, Trentham, Staffordshire, March 1970; overhauled and sold to NCB; to Norton Colliery, Staffordshire, January 1971; to Chatterley Whitfield Colliery, Tunstall, week-ending 25th September 1971; to Norton Colliery, week-ending 15th October 1971; scrapped, about April 1979.

D2865 YE 2848 1961 50D 3/70 F NFT
condemned by BR as 'surplus to requirements', 21st March 1970; to APCM, Kilvington, Nottinghamshire, September 1970; to Blue Circle, Beeston Depot, Nottingham, for storage, 1984; to The Vic Berry Company, Leicester, for scrap, December 1984; scrapped, May 1985.

D2866 YE 2849 1961 9A 2/70 P NPT
condemned by BR as 'surplus to requirements', week-ending 28th February 1970; to W.H. Arnott Young & Co Ltd, Dalmuir, Dunbartonshire; despatched from 9D Newton Heath Depot, about August 1970; to BR Glasgow Works, for repairs, April 1977; returned to W.H. Arnott Young & Co Ltd, Dalmuir; to Caledonian Railway, Brechin, 17th October 1987; to South Yorkshire Railway Preservation Society (HNRC), Meadowhall, Sheffield, 22nd January 1996; to Peak Rail, Rowsley, March 2002.

D2867 YE 2850 1961 6A 9/70 P DIANE
to Tunnel Cement, Pennyford, on hire, August 1970; to Redland Roadstone Ltd, Mountsorrel, Leicestershire, 23rd October 1970; to Barrow-upon-Soar Works, Leicestershire, late 1979; to South Yorkshire Railway Preservation Society (HNRC), Meadowhall, Sheffield, 31st March 1995; to Battlefield Line, Shackerstone, 27th July 2001.

D2868 YE 2851 1961 10D 12/69 P D2868
to Lunt, Comley and Pitt Ltd, Shut End Works, Staffordshire, mid-October 1970; to South Yorkshire Railway Preservation Society (HNRC), Meadowhall, Sheffield, 25th April 1997; to Lavender Line, Isfield, about June 2001; to Barrow Hill Engine Shed Society, Staveley, 28th January 2004; to Peak Rail, Rowsley, 13th May 2008; to Museum of Science &

Industry, Manchester, on loan, 14th November 2011; to Peak Rail, Rowsley, 1st February 2013; to Barrow Hill Engine Shed Society, Staveley, for repairs, 8th October 2018.

SECTION 10

Hunslet Engine Co Ltd built 0-4-0 diesel mechanical locomotives, numbered D2950-D2952, and introduced 1954. Fitted with a Gardner 6L3 engine developing 153bhp at 1200rpm, four speed gearbox, and driving wheels of 3ft 4in diameter. No TOPS classification.

D2950 HE 4625 1954 50D 12/67 F D4

to Llanelly Steel Co Ltd, Carmarthenshire (sold via R.E. Trem Ltd, Finningley, Doncaster and left Goole shed 29th April 1968); arrived at Llanelli, 2nd May 1968; to Thyssen Ltd, Old Castle Depot, Llanelli, for storage, May 1980; scrapped by Gwillym Jones & Son, spring 1983; the engine was retained and later used in a trawler.

SECTION 11

Andrew Barclay, Sons & Co Ltd built 0-4-0 diesel mechanical locomotives, numbered D2953-D2956, and introduced 1956. Fitted with a Gardner 6L3 engine developing 153bhp at 1200rpm, four speed gearbox, and driving wheels of 3ft 2in diameter. BR Departmental locomotive number 81 became the second D2956 after the first one was withdrawn. Later classified TOPS Class 01.

D2953 AB 395 1955 30A 6/66 P D2953

to Thames Matex Ltd, West Thurrock, Essex, June 1966 (the first BR diesel shunter sold for industrial service); to BP Refinery (Kent) Ltd, Grain, Kent, on loan, 1967; returned to Thames Matex; to Shell Mex & BP Ltd, Purfleet, on loan, at various times; to South Yorkshire Railway Preservation Society, Chapeltown, 15th December 1985; to SYRPS, Attercliffe, Sheffield, December 1986; to SYRPS, Meadowhall, Sheffield, 15th September 1988; to Peak Rail, Rowsley, March 2002.

D2956 AB 398 1956 36A 5/66 P D2956

to A. King & Sons Ltd, Norwich, July 1966; to A. King & Sons Ltd, Snailwell, Cambridgeshire, September 1981; to East Lancashire Railway, Bury, 30th July 1985.

D2956 AB 424 1958 36A 11/67 F D5

the second D2956, withdrawn by BR as Departmental locomotive number 81; to Briton Ferry Steel Co Ltd, Glamorgan, March 1968; scrapped, August 1969.

SECTION 12

Ruston & Hornsby Ltd built 0-4-0 diesel mechanical locomotives (Ruston's class 165DS), numbered D2957-D2958, and introduced 1956. Fitted with a Ruston 6VPHL engine developing 165bhp at 1250rpm, four speed gearbox, and driving wheels of 3ft 4in diameter. No TOPS classification.

D2958 RH 390777 1956 30A 1/68 F NFT
to C.F. Booth Ltd, Rotherham, May 1968 (sold via R.E. Trem Ltd, Finningley, Doncaster); used as yard shunter; scrapped, October 1984.

SECTION 13

Ruston & Hornsby Ltd built 0-6-0 diesel electric locomotives (Ruston's class LSSE), numbered D2985-D2998, and introduced 1962. Fitted with a Paxman 6RPHL engine developing 275bhp at 1360rpm, and driving wheels of 3ft 6in diameter. Later classified TOPS Class 07.

D2985 RH 480686 1962 70D 7/77 P 07001
07001 to Tilsley & Lovatt Ltd, Trentham, Staffordshire, for overhaul, 5th April 1978; to Peakstone Ltd, Holderness Limeworks, Peak Dale, Derbyshire, 30th May 1978; to South Yorkshire Railway Preservation Society (HNRC), Meadowhall, Sheffield, June 1989; to Mayer Parry Ltd, Snailwell, Cambridgeshire, on hire, 28th April 1993; returned to SYRPS, 22nd October 1997; to Barrow Hill Engine Shed Society, Staveley, 30th June 1999; to Creative Logistics, Salford, on hire, 5th March 2001; to Barrow Hill Engine Shed Society, Staveley, 20th July 2009; sold by HNRC; to Peak Rail, Rowsley, 20th December 2012.

D2986 RH 480687 1962 70D 7/77 F NFT
07002 to Powell Duffryn Fuels Ltd, NCBOE Coed Bach Disposal Point, Kidwelly, Dyfed, April 1978; scrapped on site by T. Davies of Llanelli, September 1982.

D2987 RH 480688 1962 70D 10/76 F 07003
07003 to R.E. Trem Ltd, Finningley, Doncaster, March 1977; to British Industrial Sand Ltd, Oakamoor, Staffordshire, about October 1978; scrapped, May 1985.

D2989 RH 480690 1962 70D 7/77 P 07005
07005 to Resco (Railways) Ltd, Woolwich, for overhaul (works number L106 of 1978), June 1978; to ICI Wilton Works, Middlesbrough, 18th July 1979; to Barrow Hill Engine Shed Society (HNRC), Staveley, 21st December 2000; to Battlefield Line, Shackerstone, 3rd September 2003; to Great Central Railway, Loughborough, 21st May 2008; to Boden Rail Engineering, Washwood Heath, early 2014; to Great Central Railway, Loughborough, 18th January 2018.

D2990 RH 480691 1962 70D 7/77 F NFT
07006 to Powell Duffryn Fuels Ltd, NCBOE Coed Bach Disposal Point, Kidwelly, Dyfed, April 1978; scrapped on site by T. Davis of Llanelli, October 1984.

D2991 RH 480692 1962 70D 5/73 P 07007
to Eastleigh Works, 1973; used as a stationary generator, then placed in store; to Eastleigh Railway Preservation Society, Eastleigh Works, about September 1988; to Knights Rail Services, Eastleigh Works, about March 2007; numbered 07007 (which it never carried in BR service) during major overhaul, January 2008; resumed working, 19th February 2008; to Swanage Railway, Dorset, for gala, 6th May 2008; returned to Eastleigh Works, 16th May 2008.

D2994 RH 480695 1962 70F 10/76 P 07010

07010 to Resco (Railways) Ltd, Woolwich, for overhaul; despatched from BR Eastleigh, June 1978; to Winchester & Alton Railway, New Alresford, Hampshire, August 1978; to West Somerset Railway, Minehead, 19th May 1980; to Avon Valley Railway, Bitton, Gloucestershire, 1st March 1994; to St Philip's Marsh Depot, Bristol, for tyre turning, 4th June 2014; to Avon Valley Railway, Bitton, 13th June 2014.

D2995 RH 480696 1962 70D 7/77 P D2995

07011 to Resco (Railways) Ltd, Woolwich, for overhaul (works number L105 of 1978), June 1978; to ICI Billingham Works, Stockton-on-Tees, on hire, March 1979; returned to Resco (Railways) Ltd, 12th November 1979; to ICI Wilton Works, Middlesbrough, 4th September 1980; to Hastings Diesels, St Leonards, East Sussex, 17th May 1996; to Kent & East Sussex Railway, Tenterden, Kent, 1998; to St Leonards Railway Engineering Ltd, East Sussex, about July 2000.

D2996 RH 480697 1962 70D 7/77 P 07012

07012 to Powell Duffryn Fuels Ltd, NCBOE Cwm Mawr Disposal Point, Tumble, Dyfed, April 1978; to NCBOE Coed Bach Disposal Point, Kidwelly, by 23rd April 1982; to South Yorkshire Railway Preservation Society (HNRC), Meadowhall, Sheffield, 11th December 1992; to Barrow Hill Engine Shed Society, Staveley, 28th June 1999; to Lavender Line, Isfield, for storage, June 2001; to Barrow Hill Engine Shed Society, Staveley, 18th July 2006; to Appleby Frodingham Railway Society, Scunthorpe, January 2009; to Barrow Hill Engine Shed Society, Staveley, 28th January 2016.

D2997 RH 480698 1962 70D 7/77 P 07013

07013 to Resco (Railways) Ltd, Woolwich, for overhaul (works number L101 of 1978), May 1978; to Dow Chemical Co Ltd, King's Lynn, 5th October 1978; to South Yorkshire Railway Preservation Society (HNRC), Meadowhall, Sheffield, 16th August 1994; to Barrow Hill Engine Shed Society, Staveley, 28th June 1999; to Peak Rail, Rowsley, 24th October 2003; to Barrow Hill Engine Shed Society, Staveley, 14th November 2011; sold by HNRC, May 2013; to East Lancashire Railway, Bury, 14th May 2013.

SECTION 14

British Railways built 0-6-0 diesel electric locomotives, numbered D3000-D3116, and introduced 1953. Fitted with an English Electric 6KT engine developing 350bhp at 630rpm, and driving wheels of 4ft 6in diameter. Later classified TOPS Class 08.

D3000 Derby 1952 82A 11/72 P 13000

to NCB Hafodyrynys Colliery, Pontypool, 19th March 1973; to BR Canton Depot, Cardiff, for repairs, 11th August 1975; to Hafodyrynys Colliery, Pontypool, about November 1975; to Bargoed Colliery, 17th July 1978; to BR Canton Depot, Cardiff, October 1979; to Bargoed Colliery, October 1979; to BR Canton Depot, Cardiff, June 1980; to Bargoed Colliery, July 1980; to Mountain Ash Colliery, 10th July 1981; to Mardy Colliery, 13th November 1981; to Mountain Ash Colliery, December 1981; to Mardy Colliery, May 1982; to Brighton Railway Museum, despatched on 18th March 1987; arrived at Brighton, 21st

March 1987; to South Yorkshire Railway Preservation Society, Meadowhall, Sheffield, 9th March 1993; to Barrow Hill Engine Shed Society, Staveley, 7th March 2001; to Appleby Frodingham RPS, Scunthorpe, 14th April 2008; to Peak Rail, Rowsley, 27th January 2011.

D3002 Derby 1952 82A 7/72 P 13002
to Foster Yeoman Quarries Ltd, Merehead Stone Terminal, Somerset, November 1972; at BR Westbury Depot, for repairs, February 1976; to BR Bath Road Depot, Bristol, for repairs, 13th April 1976; returned to Merehead Stone Terminal, June 1976; left Merehead and spent about two weeks stored in Westbury Yard; then moved on to Plym Valley Railway, Marsh Mills, Plymouth, 9th July 1982.

D3003 Derby 1952 82A 7/72 F MEREHEAD
to Foster Yeoman Quarries Ltd, Merehead Stone Terminal, Somerset, May 1973; to BR Bath Road Depot, Bristol, for repairs, March 1973; to BR Derby Works, for repairs, 30th June 1974; returned to Merehead Stone Terminal, late 1974; to Childrens Playground, Wanstrow, near Cranmore, Somerset, February 1982; scrapped, December 1991.

D3011 Derby 1952 70D 10/72 F LICKEY
to British Leyland Ltd, Longbridge, Birmingham, 8th January 1973; to BR Derby Works, for repair, 15th January 1976; to British Leyland Ltd, Longbridge, April 1976; to BR Tyseley Depot, Birmingham, for tyre turning, November 1981; returned to Longbridge; to Marple & Gillott Ltd, Attercliffe, Sheffield, for scrap, 6th December 1985; scrapped, December 1985.

D3014 Derby 1952 70D 10/72 P D3014 / SAMSON
to NCB Merthyr Vale Colliery, Aberfan, August 1973; despatched from BR Eastleigh; seen at Merthyr Vale Colliery on 21st August 1973; to BR Canton Depot, Cardiff, for repairs, December 1974; seen at Canton Depot, 19th December 1974; to Merthyr Vale Colliery, Aberfan, about January 1975; to BR Canton Depot, Cardiff, for repairs, 19th February 1980; to Merthyr Vale Colliery, Aberfan, 21st May 1980; to BR Canton Depot, Cardiff, for repairs, September 1980; to Merthyr Vale Colliery, Aberfan, October 1980; to BR Canton Depot, Cardiff, for repairs, 3rd October 1981; to Merthyr Vale Colliery, Aberfan, 27th December 1981; to BR Canton Depot, Cardiff, for repairs, 19th July 1985; to Merthyr Vale Colliery, Aberfan, 28th March 1986; to BR Canton Depot, Cardiff, for repairs, 5th October 1987; to Merthyr Vale Colliery, Aberfan, 30th October 1987; to Paignton & Dartmouth Steam Railway, Devon, 4th March 1989.

D3018 Derby 1953 81D 12/91 P D3018 / HAVERSHAM
08011 to Chinnor and Princes Risborough Railway, Oxfordshire, 25th April 1992.

D3019 Derby 1953 8J 6/73 P D3019
despatched from Allerton Depot, Liverpool, November 1973; to Bescot Yard, Walsall, where stored for three weeks; to Powell Duffryn Fuels Ltd, NCBOE Gwaun-cae-Gurwen Disposal Point, West Glamorgan, 13th December 1973; to BR Canton Depot, Cardiff, for repairs, 4th June 1978; to Gwaun-cae-Gurwen Disposal Point, 8th December 1978; to South Yorkshire Railway Preservation Society (HNRC), Meadowhall, Sheffield, 5th July 1990; to Battlefield Line, Shackerstone, 30th July 2001; to Tyseley Steam Depot, Birmingham, for tyre turning, November 2003; to Cambrian Railway Trust, Llynclys, Oswestry, 20th December 2003.

D3022 **Derby** **1953** **41A** **9/80** **P** **D3022**
08015 to Severn Valley Railway, Bridgnorth; despatched from BR Swindon Works, 27th May 1983; to EWS Toton Depot, for repairs, 13th April 2006; returned to Severn Valley Railway, April 2006; to St Philip's Marsh Depot, Bristol, for tyre turning, 7th July 2019; returned to Severn Valley Railway, 11th July 2019.

D3023 **Derby** **1953** **9D** **5/80** **P** **08016**
08016 to Hargreaves Industrial Services Ltd, NCBOE British Oak Disposal Point, Crigglestone, West Yorkshire; despatched from 9D Newton Heath Depot, October 1980; to South Yorkshire Railway Preservation Society, Meadowhall, Sheffield, 24th January 1992; to Peak Rail, Rowsley, 4th April 2002; to Bluebell Railway, Horsted Keynes, on hire, 27th March 2006; to Peak Rail, Rowsley, April 2008.

D3029 **Derby** **1953** **15A** **12/86** **P** **13029**
08021 to Tyseley Steam Depot, Birmingham; despatched from BR Derby, 7th May 1987; to BR Tyseley Depot, for repairs, 3rd December 1988; returned to Tyseley Steam Depot, Birmingham, December 1988.

D3030 **Derby** **1953** **41A** **3/85** **P** **LION**
08022 to Guinness Ltd, Park Royal, London; despatched from BR Swindon Works, 5th July 1985; arrived at Guinness, 20th July 1985; to BR Old Oak Common Depot, London, for repairs, April 1990; returned to Guinness Ltd; to BR Old Oak Common Depot, London, for repairs, August 1991; returned to Guinness Ltd; to RFS (Engineering) Ltd, Doncaster, for repairs, 13th December 1993; returned to Guinness Ltd, 11th January 1994; to Cholsey & Wallingford Railway, Oxfordshire, 31st August 1997.

D3038 **Derby** **1953** **9A** **12/72** **F** **2100/525**
to NCB Ashington Central Workshops; despatched from BR Newton Heath Depot, October 1973; to Bates Colliery, Blyth, March 1974; scrapped, June 1980.

D3044 **Derby** **1954** **16A** **8/74** **P** **08032 / MENDIP**
08032 to BR Derby Works, for overhaul, August 1974; emerged with plate affixed 'Overhauled and modified. Derby 1975'; to Foster Yeoman Quarries Ltd, Merehead Stone Terminal, Somerset, about 23rd February 1975; to BR Gloucester Depot, for repairs, 10th February 1980; returned to Merehead Stone Terminal, March 1980; to East Somerset Railway, Cranmore, for repairs, 10th August 2001; returned to Merehead Stone Terminal, by 18th November 2001; to East Somerset Railway, Cranmore, by 5th May 2002; returned to Merehead Stone Terminal; to East Somerset Railway, Cranmore, April 2003; returned to Merehead Stone Terminal, about February 2004; to East Somerset Railway, Cranmore, on loan, December 2004; returned to Merehead Stone Terminal; to Whatley Quarry, about March 2006; returned to Merehead Stone Terminal, 9th July 2006; to Mid Hants Railway, Ropley, on loan, 15th August 2008; to Knight's Rail Services, Eastleigh Works, 19th October 2010; to Mid Hants Railway, Ropley, 8th February 2012.

D3059 **Derby** **1954** **16F** **5/80** **P** **13059 /**
08046 **BRECHIN CITY**
to Associated British Maltsters Ltd, Airdrie; despatched from BR Derby, 21st January

1981; to BR Motherwell Depot, for repairs, 7th February 1981; returned to Associated British Maltsters Ltd; to Caledonian Railway, Brechin, December 1985.

D3067 Darlington 1953 52A 2/80 P 08054 / M414
08054 to Tilcon Ltd, Swinden Lime Works, Grassington, about August 1980; to Embsay & Bolton Abbey Railway, 27th February 2008.

D3074 Darlington 1953 40A 6/84 P 060 / UNICORN
08060 to Guinness Ltd, Park Royal, London; despatched from BR Swindon Works, 5th July 1985; arrived at Guinness, 20th July 1985; to BR Old Oak Common Depot, London, for repairs, early April 1988; returned to Guinness Ltd; to BR Old Oak Common Depot, London, for repairs, May 1989; returned to Guinness Ltd; to Cholsey & Wallingford Railway, Oxfordshire, 31st August 1997.

D3079 Darlington 1953 55B 12/84 P 13079
08064 to National Railway Museum, York, 28th October 1985; to National Railway Museum, Shildon, 18th August 2004; to National Railway Museum, York, 18th October 2005; to National Railway Museum, Shildon, 18th November 2013; to National Railway Museum, York, 8th June 2016.

D3087 Derby 1954 8F 7/73 F NFT
to CEGB Walsall Power Station, October 1973; to BR Tyseley Depot, for repairs, November 1981; returned to CEGB Walsall Power Station; scrapped on site by Thos. W. Ward Ltd, May 1983.

D3088 Derby 1954 2F 12/73 F D3088 / 2100-526
to NCB Ashington Colliery, 18th April 1974; to Ashington Workshops, 20th June 1974; to Bates Colliery, Blyth, 6th December 1974; to Lambton Engine Works, Philadelphia, 20th April 1979; to Bates Colliery, Blyth, 6th September 1979; to Lambton Engine Works, Philadelphia, 29th June 1981; to Bates Colliery, Blyth, 15th February 1983; scrapped on site by C.H. Newton & Co Ltd of Durham, November 1985.

D3099 Derby 1955 73F 10/72 F NFT
to Shipbreaking (Queenborough) Ltd, Kent, July 1973; used for spares; remains scrapped (possibly at the wharf), circa 1980.

D3101 Derby 1955 73F 5/72 P 13101
to ARC (East Midlands) Ltd, Loughborough, February 1973; to Great Central Railway, Loughborough, 14th December 1984.

D3102 Derby 1955 31A 11/77 P 007 / JAMES
08077 to Wiggins Teape & Co Ltd, Fort William, December 1978; to BR Eastfield Depot, Glasgow, for repairs, May 1979; returned to Wiggins Teape; to BR Glasgow Works, for repairs, June 1981; returned to Wiggins Teape; to BR Eastfield Depot, for repairs, 16th March 1984; returned to Wiggins Teape, 24th March 1984; sold to RFS (Engineering) Ltd, Kilnhurst, 23rd November 1990; to BREL, York, on hire, 16th December 1991; returned to RFS (Engineering) Ltd, 17th November 1992; to Roche Products, Dalry, on hire, 17th November 1992; to RFS (Engineering) Ltd, Doncaster, about March 1993; to Teesbulk Handling, Middlesbrough, on hire, 2nd June 1994; returned to RFS

(Engineering) Ltd, 13th September 1994; to Eastleigh, for storage, 10th December 1996; sold to Freightliner, December 1996, and initially based at Southampton Docks.

D3110 Derby 1955 52A 3/86 F 08085
08085 to RFS (Engineering) Ltd, Doncaster, July 1988; to RFS (Engineering) Ltd, Kilnhurst, July 1989; returned to RFS (Engineering) Ltd, Doncaster, by 4th March 1990; used for spares; remains to C.F. Booth Ltd, Rotherham, for scrap, 24th March 1993; scrapped, 16th April 1993.

SECTIONS 15/16

Two variants of the popular shunter: Firstly British Railways built 0-6-0 diesel electric locomotives, numbered D3127-D3136, D3167-D3438, D3454-D3472, D3503-D3611, D3652-D3664, D3672-D3718, D3722-D4048, D4095-D4098, and D4115-D4192, introduced 1953. (Locomotives D3117-D3126 and D3152-D3166 were unclassified.) Fitted with an English Electric 6KT engine developing 350bhp at 680rpm, and driving wheels of 4ft 6in diameter. Later classified TOPS Class 08. Secondly British Railways built 0-6-0 diesel electric locomotives, numbered D3137-D3151, D3439-D3453, D3473-D3502, D3612-D3651 and D4049-D4094, introduced 1955. Fitted with a Lister-Blackstone ER6T engine developing 350bhp at 750rpm, and driving wheels of 4ft 6in diameter. Later classified TOPS Class 10.

D3167 Derby 1955 36A 4/88 P D3167
08102 purchased by Lincoln City Council, 23rd March 1988; displayed at Central Station, Lincoln, from 23rd August 1988; to BR Doncaster Works, for overhaul, 12th October 1988; returned to display at Central Station, Lincoln, 5th April 1989; to Lincolnshire Wolds Railway, Ludborough, 8th May 1994.

D3174 Derby 1955 31A 7/84 P D3174 /
08108 DOVER CASTLE
to Dower Wood & Co Ltd, Newmarket, Suffolk, 1st August 1984; to East Kent Railway, Shepherdswell, 3rd August 1991; to Kent & East Sussex Railway, Tenterden, 13th October 1992.

D3179 Derby 1955 86A 3/84 F 08113 / HO17
08113 to Powell Duffryn Fuels Ltd, NCBOE, Gwaun-cae-Gurwen; despatched from Canton Depot, Cardiff, 2nd August 1984; arrived on 6th August 1984; sold to RMS Locotec, Dewsbury, 21st September 1995; to RMS Locotec, Wakefield, April 2006; to Morley Waste Traders Ltd, Leeds, February 2007; scrapped, May 2007.

D3180 Derby 1955 36A 11/83 P 13180
08114 to Gloucestershire Warwickshire Railway, Toddington; despatched from BR Swindon Works, 2nd October 1984; arrived at GWR, 3rd October 1984; to Swindon & Cricklade Railway, 14th November 1987; to Great Central Railway, Loughborough, 22nd January 1991; to Great Central Railway, Ruddington, Nottingham, 25th February 1997.

D3183 Derby 1955 82C 12/72 F D3183
to NCB Merthyr Vale Colliery, Aberfan, 19th March 1973; to BR Canton Depot, Cardiff,

for repairs, May 1975; seen at Canton Depot, 29th May 1975; to Merthyr Vale Colliery, Aberfan, June 1975; to BR Canton Depot, Cardiff, for repairs, 11th September 1980; to Merthyr Vale Colliery, Aberfan, 19th December 1980; to BR Canton Depot, Cardiff, for repairs, 19th July 1985; to Merthyr Vale Colliery, Aberfan, September 1986; scrapped by W. Phillips Ltd of Llanelli, December 1987.

D3190 Derby 1955 5A 3/84 P GEORGE MASON
08123 to Cholsey & Wallingford Railway, Oxfordshire; despatched from BR Swindon Works, 7th June 1985.

D3201 Derby 1955 40A 9/80 P 13201
08133 to Sheerness Steel Co Ltd, Sheerness, Kent; despatched from BR Swindon Works, 6th October 1981; to RFS (Engineering) Ltd, Kilnhurst, for repairs, 29th March 1991; returned to Sheerness Steel Co Ltd, 1st May 1991; to South Yorkshire Railway Preservation Society, Meadowhall, Sheffield, 30th August 1995; to Barrow Hill Engine Shed Society, Staveley, 19th December 2000; to Severn Valley Railway, Bridgnorth, 17th April 2002.

D3225 Darlington 1955 73F 4/77 F 009
08157 to Independent Sea Terminals, Ridham Dock, Kent; despatched from BR Eastleigh Works, 28th July 1977; to RFS (Engineering) Ltd, Doncaster, 30th April 1993; to European Metal Recycling, Attercliffe, Sheffield, 28th May 1996; scrapped, 20th June 1996.

D3232 Darlington 1956 52A 3/86 P 08164 / PRUDENCE
08164 to RFS (Engineering) Ltd, Doncaster; despatched from BR Tyne Yard, July 1988; to RFS (Engineering) Ltd, Kilnhurst, for weighing, 15th January 1991; returned to RFS (Engineering) Ltd, Doncaster, same day; to East Lancashire Railway, Bury, 28th January 1998.

D3236 Darlington 1956 55B 3/88 P 13236
08168 to ABB Transportation, York, April 1989; locomotive included in sale when BREL works privatised; sold at public auction, 25th June 1996; to Battlefield Line, Shackerstone, 12th July 1996; to Fragonset, Derby, 1st July 1999; to Alstom Transport, Old Dalby Test Centre, 10th December 2001; to Battlefield Line, Shackerstone, 22nd July 2004; to Bluebell Railway, Horsted Keynes, on hire, 30th April 2008; to Nemesis Rail, Burton upon Trent, 27th June 2014; to Epping Ongar Railway, Essex, for gala, 21st April 2016; to Nemesis Rail, Burton upon Trent, 30th April 2016.

D3238 Darlington 1956 52A 3/86 F 08170
08170 to RFS (Engineering) Ltd, Doncaster; despatched from BR Tyne Yard, July 1988; to RFS (Engineering) Ltd, Kilnhurst, 15th July 1989; used for spares; remains to C.F. Booth Ltd, Rotherham, 11th December 1991; scrapped, December 1991.

D3245 Derby 1956 55B 10/88 F D3245
08177 to ABB Transportation, Crewe, April 1989; locomotive included in sale when BREL works privatised; scrapped on site by M.R.J. Phillips (Metals) Ltd, August 1996.

D3255 Derby 1956 85B 12/72 P D3255
to NCB Blaenavon Colliery, 19th March 1973; to Bargoed Colliery, by 22nd March 1973;

to BR Canton Depot, Cardiff, for tyre turning, September 1974; seen at Canton Depot, 26th September 1974; to Bargoed Colliery; to BR Canton Depot, Cardiff, for repairs, 28th December 1975; to Bargoed Colliery, about April 1976; to BR Canton Depot, Cardiff, August 1976; to Bargoed Colliery, 1976; to BR Canton Depot, Cardiff, for repairs, 18th April 1977; to Bargoed Colliery, August 1977; to BR Canton Depot, Cardiff, March 1978; to Bargoed Colliery, about May 1978; to BR Canton Depot, Cardiff, 23rd May 1979; to Bargoed Colliery, about August 1979; to BR Canton Depot, Cardiff, 6th April 1981; to Mardy Colliery, 27th May 1981; to Mountain Ash Colliery, 28th December 1981; to Mardy Colliery, 27th May 1982; to Brighton Railway Museum, about May 1987; to Colne Valley Railway, Castle Hedingham, 9th September 2008; to Tim Ackerley, c/o T.W.S. Welders, North Side Works, Leavening, Malton, 27th August 2009.

D3261 Derby 1956 86A 12/72 P D3261

to NCB Tower Colliery, Hirwaun, Glamorgan, July 1973; to BR Canton Depot, Cardiff, for repairs, September 1975; to Tower Colliery, Hirwaun, October 1975; to BR Canton Depot, Cardiff, 27th October 1977; to Tower Colliery, Hirwaun, 2nd November 1977; to BR Canton Depot, Cardiff, 28th October 1978; to BR Swindon Works, April 1979; to Tower Colliery, Hirwaun, 29th November 1979; to Brighton Railway Museum, 11th December 1988; to Swindon & Cricklade Railway, about March 1996.

D3265 Derby 1956 86A 9/83 P 13265 / MARK

08195 to Llangollen Railway; despatched from BR Swindon Works, 25th March 1986.

D3272 Derby 1956 86A 5/89 P 08202

08202 to G.G. Papworth Ltd, Ely, Cambridgeshire, 10th March 1990; to The Potter Group Ltd, Knowsley, Merseyside, 9th January 2001; to Gloucestershire Warwickshire Railway, Toddington, for gala and then storage, 7th July 2010; to GBRf, Celsa, Cardiff, on hire, late January 2011; to The Potter Group Ltd, Ely, 4th May 2011; to Chasewater Railway, Staffordshire, 15th January 2014; to Avon Valley Railway, Bitton, 30th June 2015; to Knorr Bremse Rail Services Ltd, Wolverton Works, on hire, 25th January 2018; to Avon Valley Railway, Bitton, 16th May 2018.

D3286 Derby 1956 16C 11/80 F 08216

08216 to Sheerness Steel Co Ltd, Sheerness, Kent; despatched from BR Swindon Works, 22nd April 1983; arrived at Sheerness, 12th May 1983; to RFS (Engineering) Ltd, Kilnhurst, for overhaul, 21st December 1989; returned to Sheerness Steel Co Ltd, 5th May 1990; dismantled by June 1995; to South Yorkshire Railway Preservation Society (HNRC), Meadowhall, Sheffield, 2nd April 1996; to Barrow Hill Engine Shed Society, Staveley, 10th April 2001; scrapped, 19th April 2001.

D3290 Derby 1956 5A 3/86 P 08220

08220 to William Smith Ltd, Wakefield, West Yorkshire; despatched from BR Chester Depot, 11th July 1988; to Steamtown, Carnforth, Lancashire, 12th July 1990; to Wrenbury Station, Cheshire, April 2006; to Great Central Railway, Ruddington, Nottingham, October 2008; sold to Railway Support Services, Wishaw, Warwickshire; to Electro-Motive Diesel, Longport, Staffordshire, on hire, week-ending 9th August 2014.

D3308 Darlington 1956 85B 3/84 P 13308 / CHARLIE

08238 sold to Forest Free Mining, Tetbury, 1984; intended for Parkend Mine, Forest of Dean, but stored at BR Gloucester Depot until November 1988; to BR Swindon Works, 11th November 1988; to Swindon Heritage Centre, 29th November 1988; to Dean Forest Railway, Lydney, January 1993; to RFS (Engineering) Ltd, Doncaster, for repairs, December 1997; to Dean Forest Railway, Lydney, 1998.

D3309 Darlington 1956 55H 7/84 F 3309
08239 to C.F. Booth Ltd, Rotherham, 1995; resold to South Yorkshire Railway Preservation Society (HNRC), Meadowhall, Sheffield, 12th November 1998; to European Metal Recycling, Kingsbury, about June 2001; scrapped, October 2005.

D3336 Darlington 1957 41J 3/85 P 08266
08266 to Keighley & Worth Valley Railway, Haworth; despatched from BR Swindon Works, 21st November 1985.

D3342 Derby 1957 31B 7/87 F 08272
08272 to RFS (Engineering) Ltd, Doncaster, South Yorkshire, April 1988; used for spares; remains scrapped, January 1991.

D3358 Derby 1957 82C 1/83 P D3358
08288 to Mid Hants Railway, Ropley; despatched from BR Swindon Works, 13th September 1984; arrived at Ropley, 1st November 1984; to Wabtec, Doncaster, for repairs, 20th August 2002; returned to Mid Hants Railway, Ropley, October 2002.

D3362 Derby 1957 65A 5/84 F 08292
08292 to Deanside Transit Ltd, Glasgow, July 1984; scrapped on site, November 1994.

D3366 Derby 1957 55B 10/88 F 08296
08296 to ABB Transportation, Crewe, April 1989; locomotive included in sale when BREL works privatised; between July 1991 and October 1992 it was given identity 08787; to C. F. Booth Ltd, Rotherham, for scrap, 16th February 1994; scrapped 13th May 1994.

D3378 Derby 1957 36A 11/92 P 08308 / 23
08308 to South Yorkshire Railway Preservation Society, Meadowhall, Sheffield, 15th July 1993; purchased by RT Rail, Crewe, 1st May 1998; to RFS, Doncaster, for repairs, 21st January 1999; to Scot Rail, Inverness, on hire, 7th September 1999; to Wabtec, Doncaster, for repairs, 16th November 2005; to Scot Rail, Inverness, on hire, 4th May 2006; RT Rail acquired by RMS Locotec, 8th November 2007; to Castle Cement Works, Ketton, on hire, 30th August 2013; to RMS Locotec, Weardale Railway, Wolsingham, 16th January 2014; to PD Ports, Tees Dock, on hire, 1st August 2014; to RMS Locotec, Weardale Railway, Wolsingham, 3rd October 2018.

D3390 Derby 1957 16A 12/82 F P400D / SUSAN
08320 sold to Forest Free Mining, Tetbury; despatched from 16A Toton Depot, 2nd October 1984; intended for Parkend Mine, Forest of Dean, but stored at Gloucester Depot from 5th October 1984 until December 1988; to English China Clays Ltd, Blackpool Driers, Burngullow, Cornwall, 2nd December 1988; to Imerys, Rocks Driers, Bugle, December 2008; to European Metal Recycling, Kingsbury, for scrap, 9th September 2010; scrapped October 2010.

D3401 Derby 1957 36A 4/88 P 08331 / TERENCE

08331 to RFS (Engineering) Ltd, Doncaster, April 1988; to RFS (Engineering) Ltd, Kilnhurst, 5th August 1991; to Inco Europe, Clydach, Swansea, on hire, 17th September 1991; returned to RFS (Engineering) Ltd, Kilnhurst, 6th November 1991; to Trans-Manche Link, Channel Tunnel contract (number 95), on hire, 26th September 1992; to RFS (Engineering) Ltd, Kilnhurst, 6th February 1993; to Flixborough Wharf Ltd, Flixborough, Scunthorpe, on hire, 8th March 1993; to Allied Steel & Wire Ltd, Cardiff, on hire, 24th September 1994; returned to RFS (Engineering) Ltd, Doncaster, 27th November 1994; to Great North Eastern Railway, Craigentinny Depot, Edinburgh, on hire, 30th January 1997; returned to RFS (Engineering) Ltd, Doncaster, by February 1998; to Hays Chemicals, Sandbach, on hire, August 1998; returned to RFS (Engineering) Ltd, September 1998; to GNER, Craigentinny Depot, Edinburgh, on hire, 12th October 1998; to Wabtec, Doncaster, for repairs, December 2000; returned to Craigentinny Depot, Edinburgh; to Wabtec, Doncaster, 11th August 2005; purchased by RT Rail, Crewe, 7th May 2006; to Embsay & Bolton Abbey Railway, for testing, May 2007; purchased by RMS Locotec, November 2007; to Wabtec, Doncaster, by 5th December 2007; to LaFarge Aggregates, Barrow upon Soar, on hire, 13th December 2007; to Midland Railway, Butterley, for storage, by 5th July 2008; to Cemex Rail Products, Washwood Heath, on hire, August 2010; to Boden Rail Engineering, Washwood Heath, August 2010; to Wabtec, Doncaster, for repairs, early August 2011; to Cemex Rail Products, Washwood Heath, on hire, September 2011; to Midland Railway, Butterley, Derbyshire, early February 2012.

D3405 Derby 1957 41A 1/87 F 08335

08335 to Thomas Hill (Rotherham) Ltd, Kilnhurst, South Yorkshire, 20th November 1987; to RFS (Engineering) Ltd, Kilnhurst, July 1989; scrapped on site by C.F. Booth Ltd, Rotherham, 26th July 1989.

D3407 Derby 1957 36A 2/87 F 08337

08337 to RFS (Engineering) Ltd, Doncaster, April 1988; used for spares; remains scrapped, January 1989.

D3415 Derby 1958 67C 11/83 F RUSSELL

08345 to Deanside Transit Ltd, Glasgow, May 1985; to Harry Needle Railroad Company, Long Marston, 21st June 2007; to C.F. Booth Ltd, Rotherham, for scrap, 4th November 2009; scrapped 13th November 2009.

D3420 Crewe 1957 86A 1/84 F D3420

08350 to Churnet Valley Railway, Cheddleton; despatched from BR Swindon Works, 17th September 1984; arrived at Cheddleton, 18th September 1984; to L&NWR, Carriage Works, Crewe, July 2004; to Midland Railway, Butterley, 5th September 2007; to Heanor Haulage, Langley Mill, for storage, early September 2008; to Ron Hull Ltd, Rotherham, for scrap, week-ending 19th September 2008; scrapped, March 2010.

D3429 Crewe 1958 86A 1/84 P D3429

08359 to Churnet Valley Railway, Cheddleton; despatched from BR Swindon Works, 19th September 1984; arrived at Cheddleton, 20th September 1984; to Peak Rail, Buxton, 9th January 1987; to Peak Rail, Darley Dale, December 1989; to Battlefield Line, Shackerstone, 16th October 1996; to Tyseley Steam Depot, Birmingham, 29th June 1999;

to Northampton & Lamport Railway, Chapel Brampton, 11th August 2005; to Telford Steam Railway, Shropshire, 20th January 2007; to Chasewater Railway, Staffordshire, on loan, 19th July 2010; to Telford Steam Railway, Shropshire, 25th November 2016; to Chasewater Railway, Staffordshire, 17th January 2017; to Telford Steam Railway, Shropshire, 12th November 2018; to Chasewater Railway, Staffordshire, 4th April 2019.

D3452 Darlington 1957 16A 6/68 P D3452
to ECC Ports Ltd, Fowey, Cornwall, September 1968; to Bodmin & Wenford Railway, Cornwall, 5th March 1989.

D3460 Darlington 1957 86A 11/91 P 08375 / 21
08375 to Railway Age, Crewe, 13th August 1993; to L&NWR, Carriage Works, Crewe, about January 1998; purchased by RT Rail, Crewe; to RMS Locotec, Dewsbury, for air-brake equipping, 25th March 1998; to Direct Rail Services, Sellafield, on hire, about August 1998; to RFS (Engineering) Ltd, Doncaster, for repairs, about January 1999; to Port of Felixstowe, on hire, 2nd November 1999; to Freightliner, Ipswich, on hire, by 21st November 1999; to Freightliner, Port of Felixstowe, on hire, by 10th July 2001; to Wabtec, Doncaster, for repairs, by 7th October 2001; to Flixborough Wharf Ltd, Flixborough, Scunthorpe, on hire, January 2002; to Freightliner, Port of Felixstowe, on hire, by 16th February 2002; to Wabtec, Doncaster, May 2002; to Hays Chemicals, Sandbach, on hire, 5th July 2002; returned to Wabtec, Doncaster, for repairs, 2nd December 2002; to Hays Chemicals, Sandbach, on hire, about January 2003; to Wabtec, Doncaster, March 2003; to Port of Felixstowe, on hire, 4th June 2003; to Wabtec, Doncaster, for repairs, by 6th October 2003; to Alstom Transport, Eastleigh Works, on hire, 20th November 2003; to Hays Chemicals, Sandbach, on hire, September 2004; to Wabtec, Doncaster, for repairs, 2005; returned to Hays, on hire, 15th September 2005; to PD Ports, Tees Dock, on hire, 15th October 2005; to Wabtec, Doncaster, for repairs, 15th February 2006; to Manchester Ship Canal, Trafford Park, Manchester, on hire, about May 2006; to Wabtec, Doncaster, December 2006; to DHL, ProLogis Park Industrial Estate, Coventry, on hire, 5th February 2007; to Corus, Trostre Works, Llanelli, on hire, 18th December 2009; to Castle Cement Works, Ketton, on hire, November 2010; to PD Ports, Tees Dock, on hire, 28th February 2011; to Tata Steel, Shotton, on hire, week commencing 25th March 2013; to Castle Cement Works, Ketton, on hire, 15th January 2014; to RMS Locotec, Weardale Railway, Wolsingham, for overhaul and repaint, 11th May 2019; sold to Port of Boston, June 2019; to Port of Boston, 11th July 2019.

D3462 Darlington 1957 84A 6/83 P D3462
08377 to Dean Forest Railway, Lydney; despatched from BR Swindon Works, 19th March 1986; to Rail & Marine Engineering, Thingley Junction, Chippenham, for storage, 1995; to West Somerset Railway, Minehead, 23rd April 1996; to Mid Hants Railway, Ropley, 26th March 2013.

D3476 Darlington 1957 16A 6/68 F D3476
to ECC Ports Ltd, Fowey, Cornwall, 6th September 1968; to storage in the Midlands, 5th March 1989; to South Yorkshire Railway Preservation Society (HNRC), Meadowhall, Sheffield, 2nd October 1989; to Colne Valley Railway, Castle Hedingham, 1st December 2000; to hauliers yard (possibly Wishaw) where used for spares, 26th February 2009; remains to T.J. Thomson Ltd, Stockton-on-Tees, 3rd March 2009; scrapped March 2009.

D3489 Darlington 1958 16A 4/68 P D3489
to Felixstowe Dock & Railway Co Ltd, Suffolk, August 1968; seen re-painted dark blue and fitted with air brakes, 27th September 1972; to BR Swindon Works, for repairs, 30th January 1980; to Port of Felixstowe, 19th May 1980; to BR Stratford Depot, for repairs, 11th October 1984; to Port of Felixstowe, December 1984; to Wilmott Bros, Ilkeston, for repairs, 27th July 1990; to Port of Felixstowe, 18th September 1990; to Spa Valley Railway, Tunbridge Wells, 17th August 2001.

D3497 Doncaster 1957 16B 4/68 F D3497
to ECC Ports Ltd, Fowey, Cornwall, 21st August 1968; dismantled (as a source of spares for D3452 and D3476), 1988; remains scrapped, February 1990.

D3503 Derby 1958 40B 6/96 F 08388
08388 sold to Mike Darnall, Newton Heath, Manchester, about 1999; to Wabtec, Doncaster, for repairs, 1999; returned to Mike Darnall about July 2000; to European Metal Recycling, Kingsbury, for scrap, 15th July 2010; scrapped, early October 2010.

D3504 Derby 1958 2F 9/09 P 08389
08389 sold to Harry Needle Railroad Company; to Nemesis Rail, Burton upon Trent; despatched from EWS Toton Depot, 5th October 2011; to Barrow Hill Engine Shed Society, Staveley, 28th February 2014; to Celsa, Cardiff, on hire, 28th January 2016; to Traditional Traction, Wishaw, Warwickshire, for repairs, 28th May 2016 and still there 9th July 2016; to Celsa, Cardiff, on hire, by 27th July 2016; to Barrow Hill Engine Shed, Staveley, for repairs, 26th November 2016; to Midland Road Depot, Leeds, for tyre turning, 24th January 2017; returned to Barrow Hill Engine Shed Society, Staveley; to Celsa, Cardiff, on hire, 10th February 2017; returned to Barrow Hill Engine Shed Society, Staveley, 27th June 2017; to Celsa, Cardiff, on hire, 7th July 2017; to Barrow Hill Engine Shed Society, Staveley, for repairs, 6th June 2019.

D3505 Derby 1958 86A 3/93 F 08390
08390 to South Yorkshire Railway Preservation Society (HNRC), Meadowhall, Sheffield; despatched from Adtranz, Crewe, 4th April 1997; used for spares; remains to Barrow Hill Engine Shed Society, Staveley, 6th March 2001; scrapped at Barrow Hill by HNRC, March 2004.

D3508 Derby 1958 16A 4/10 F 08393
08393 to European Metal Recycling, Kingsbury, 4th August 2011; to LH Group, Barton under Needwood, 11th January 2012; scrapped on site, early February 2016.

D3513 Derby 1958 82A 7/85 F 402D / ANNABEL
08398 to ECC Ports Ltd, Fowey, Cornwall, 13th December 1985; to BR Laira Depot, Plymouth, for repairs, summer 1988; to ECC Marsh Mills, Devon, July 1988; to ECC Blackpool Driers, Burngullow, by 20th August 1992; to BR Laira Depot, Plymouth, for repairs, 29th September 1992; returned to ECC Blackpool Driers; to ECC Rocks Driers, near Bugle, November 1992; to European Metal Recycling, Kingsbury, for scrap, 9th September 2010; scrapped, January 2011.

D3516 Derby 1958 40B 2/04 P 08401
08401 purchased by Hunslet Engine Company and moved to LH Group, Barton

under Needwood, 27th January 2011; repaired; to GBRf, Whitemoor Yard, March, on hire, 18th March 2011; to GBRf, Cardiff, on hire, 6th April 2011; to GBRf, Whitemoor Yard, March, on hire, 26th May 2011; to Celsa, Cardiff, on hire, October 2011; to GBRf, Whitemoor Yard, March, on hire, 8th December 2011; to LH Group, Barton under Needwood, 16th January 2013; to Cleveland Potash, Boulby Mine, on hire, 31st January 2013; to Celsa, Cardiff, on hire, 24th October 2013; to LH Group, Barton under Needwood, 5th March 2015; to Wabtec, Doncaster, 19th March 2015; to LH Group, Barton under Needwood, by 21st August 2015; to Bounds Green Depot, London, on hire, 24th August 2015; to LH Group, Barton under Needwood, by 21st August 2015; to Bounds Green Depot, London, on hire, 24th August 2015; to LH Group, Barton under Needwood, by 17th October 2015; to Midland Road Depot, Leeds, for tyre turning, 2nd August 2016; to LH Group, Barton under Needwood, mid-August 2016; to Hams Hall Rail Freight Terminal, Coleshill, on hire, 22nd August 2016.

D3520 Derby 1958 5A ? P 08405
08405 to Railway Support Services, Wishaw, Warwickshire; despatched from DB Cargo, Crewe Electric Depot, January 2017; to East Midlands Trains, Neville Hill Depot, Leeds, on hire, 23rd January 2017.

D3526 Derby 1958 66B 4/04 P 08411
08411 to Traditional Traction, Wishaw, 2007; to Colne Valley Railway, Castle Hedingham, for storage, 29th March 2007; to Railway Support Services, Wishaw, Warwickshire, 12th June 2015; used for spares.

D3528 Derby 1958 41A 12/96 F 08413
08413 to C.F. Booth Ltd, Rotherham; despatched from Wabtec, Doncaster, 13th June 2000; sold to RMS Locotec, Dewsbury, 3rd June 2004; to Morley Waste Traders Ltd, Leeds, for scrap, February 2007; scrapped, February 2007.

D3529 Derby 1958 81A 7/99 F 08414
08414 to Traditional Traction, Wishaw, Warwickshire; despatched from EWS Toton Depot, 9th March 2007; to European Metal Recycling, Kingsbury, for scrap, 16th August 2007; scrapped, late August 2007.

D3530 Derby 1958 8F 10/96 F 08415
08415 to RFS (Engineering) Ltd, Doncaster, 30th September 1996; used for spares; remains to European Metal Recycling, Attercliffe, Sheffield, for scrap, 5th September 1997; scrapped, 1997.

D3531 Derby 1958 16A 2/92 F 08416
08416 to RFS (Engineering) Ltd, Kilnhurst, 10th April 1992; used for spares; remains scrapped on site at Kilnhurst, August 1993.

D3532 Derby 1958 ? ? P 08417
08417 owned by Serco Railtest and allocated to Etches Park Depot, Derby; to Foster Yeoman Quarries Ltd, Merehead Stone Terminal, on hire, 17th June 1999; to Whatley Quarry, on hire, by 5th December 1999; returned to Serco Railtest, Derby, 30th March 2000; to Merehead Stone Terminal, on hire, by 5th April 2000; to Railway Technical Centre, Derby, 9th April 2002.

D3533 Derby 1958 16A 2/04 P 08418

08418 to West Coast Railway Company, Carnforth; despatched from Bescot Depot, 4th August 2010.

D3534 Derby 1958 12B 4/93 F 08419

08419 sold to ABB Transportation, Crewe, about April 1994; to Mike Darnall, Newton Heath, Manchester, and moved to Bombardier Transportation, Doncaster Works, for storage, 23rd November 2000; to C.F. Booth Ltd, Rotherham, for scrap, 24th November 2004; scrapped, November 2004.

D3536 Derby 1958 ? ? P 09201

09201 former number 08421 to 22nd September 1992; purchased by Harry Needle Railroad Company, 2015; to Hope Cement Works, Derbyshire, on hire, 14th September 2015.

D3538 Derby 1958 8F 11/88 P 08423 / 2 / HO11

08423 to Trafford Park Estates Ltd, Manchester, 21st November 1988; to MoD Kineton, for trials, 1st August 1994; sold to RMS Locotec, Dewsbury, 8th August 1994; to Mobil Oil Co Ltd, Coryton Bulk Terminal, Essex, on hire, about July 1995; to Flixborough Wharf Ltd, Flixborough, Scunthorpe, on hire, mid-June 1998; returned to RMS Locotec, 3rd July 2003; to Flixborough Wharf Ltd, Flixborough, Scunthorpe, on hire, 18th February 2004; suffered fire damage, 11th September 2006; to RMS Locotec, Wakefield, for repairs, 10th October 2006; to PD Ports, Tees Dock, on hire, week-ending 23rd February 2007; to RMS Locotec, Weardale Railway, Wolsingham, 1st August 2014; to PD Ports, Tees Dock, on hire, 27th August 2014.

D3543 Derby 1958 ? ? P 08428

08428 to Harry Needle Railroad Company; despatched from DB Cargo, Warrington Depot, January 2017; to Celsa, Cardiff, on hire, 24th January 2017; to Moveright International, Wishaw, Warwickshire, 4th March 2017; to Barrow Hill Engine Shed Society, Staveley, 14th March 2017.

D3551 Derby 1958 36A 1/92 P 08436

08436 to South Yorkshire Railway Preservation Society, Meadowhall, Sheffield, late May 1993; purchased by RT Rail, Crewe, about April 1998; to RMS Locotec, Dewsbury, for overhaul, 1st May 1998; to Keighley & Worth Valley Railway, Haworth, for painting, 1998; to Hays Chemicals, Sandbach, on hire, 8th January 1999; to Railway Age, Crewe, for repairs, 17th December 2000; to Hays Chemicals, Sandbach, on hire, 22nd December 2000; to Wabtec, Doncaster, for repair, 26th September 2001; to Hays Chemicals, Sandbach, on hire, 9th February 2002; to RT Rail, Crewe, November 2002; to Hays Chemicals, Sandbach, on hire, March 2003; to Wabtec, Doncaster, by July 2003; to Swanage Railway, Dorset, August 2004.

D3556 Derby 1958 66B 2/04 P 08441

08441 to Traditional Traction, Wishaw, Warwickshire; despatched from EWS Motherwell Depot, 14th May 2007; to Colne Valley Railway, Castle Hedingham, June 2007; to LH Group, Barton under Needwood, for repairs, 16th February 2010; to Colne Valley Railway, Castle Hedingham, November 2010; to Port of Felixstowe, on hire, 7th

February 2011; to Railway Support Services, Wishaw, Warwickshire, 24th April 2014; to Chasewater Railway, Staffordshire, for testing, 19th December 2015; to Bounds Green Depot, London, on hire, 3rd February 2016.

D3557 Derby 1958 81A 1/04 P 08442 / 0042
08442 to London & North Western Railway Company, Traction and Rolling Stock Depot, Eastleigh, Hampshire (ex DB Schenker, with site), 21st April 2011.

D3558 Derby 1958 62A 7/85 P D3558
08443 to Scottish Grain Distillers, Cambus Distillery, Alloa; despatched from BR Grangemouth Depot, 14th January 1986; to Scottish Railway Preservation Society, Bo'ness, 28th June 1993.

D3559 Derby 1958 86A 11/86 P 08444
08444 to Bodmin & Wenford Railway, Cornwall, 27th March 1987.

D3560 Derby 1958 40B 11/95 P 08445
08445 purchased by Mike Darnall, Newton Heath, Manchester, about 1999; to Wabtec, Doncaster, for repairs, 1999; returned to Mike Darnall, 17th July 2000; to East Lancashire Railway, Bury, by 14th September 2001; to Carillion Construction Ltd, Manchester Metrolink upgrade, on hire, 26th May 2007; to former Corus Works, Castleton, for storage, from about 31st August 2007; to Castle Cement Works, Ketton, on hire, 18th February 2009; to Corus, Shotton Steelworks, on hire, 22nd July 2009; purchased by LH Group, Barton under Needwood, March 2011; to LH Group, Barton under Needwood, for repairs, 27th March 2011; to GBRf, Trafford Park, Manchester, on hire, April 2011; to Daventry International Rail Freight Terminal, on hire, week-ending 24th June 2011.

D3562 Derby 1958 12B 11/94 P 08447
08447 to Deanside Transit Ltd, Glasgow, about May 1995; to Harry Needle Railroad Company, Long Marston, for overhaul, 12th June 2007; to Deanside Transit Ltd, Glasgow, 13th June 2008; to John G. Russell (Transport) Ltd, Hillington, Glasgow, for storage.

D3566 Derby 1958 ? ? P 08451
08451 acquired by Alstom Transport, Longsight Depot, Manchester, 2007; to Alstom Transport, Wembley Depot, London, 24th January 2012; to Alstom Transport, Longsight Depot, Manchester, 14th March 2012; to Alstom Transport, Polmadie Depot, Glasgow, by 22nd April 2016; to Alstom Transport, Edge Hill Depot, Liverpool, week-ending 23rd June 2017; to Alstom Transport, Longsight Depot, Manchester, by 31st October 2017.

D3569 Derby 1958 ? ? P 08454
08454 acquired by Alstom Transport, Wembley Depot, London, 2007; to Alstom Transport, Longsight Depot, Manchester, 5th March 2012; to Alstom Transport, Oxley Depot, Wolverhampton, by 28th June 2014; to Alstom Transport, Edge Hill Depot, Liverpool, by 23rd February 2015; to Arlington Fleet Services, Eastleigh Works, for overhaul, by 27th September 2016; to Alstom Transport, Edge Hill Depot, Liverpool, by 14th December 2016; to Alstom Transport, Technology Centre, Widnes, week-ending 23rd June 2017.

D3575 Crewe 1958 8J 4/04 P 08460

08460 sold to Traditional Traction, Wishaw; moved to Colne Valley Railway, Castle Hedingham; despatched from EWS Allerton Depot, 29th March 2007; to Port of Felixstowe, on hire, 5th November 2009; to Colne Valley Railway, Castle Hedingham, for repairs,13th January 2011; to Traditional Traction, Wishaw, June 2014; to Axiom Rail, Wagon Works, Stoke on Trent, on hire, about September 2014; to Traditional Traction, Wishaw, for repairs, 6th December 2016; to DB Cargo Maintenance Ltd, Wheildon Road Wagon Works, Stoke on Trent, on hire, 5th January 2017; to Railway Support Services, Wishaw, 8th February 2019; to GBRf, East Yard, Eastleigh, on hire, 28th March 2019.

D3577 Crewe 1958 16A 3/15 P 08994

08994 cut down cab locomotive; formerly numbered 08462 until 4th September 1987; sold to Harry Needle Railroad Company, August 2015; to Nemesis Rail, Burton upon Trent; despatched from DB Schenker, Toton Depot, 14th September 2015.

D3585 Crewe 1958 5A 3/86 F 08470

08470 to ABB Transportation, Crewe, April 1989; locomotive included in sale when BREL works privatised; scrapped on site by M.R.J. Phillips (Metals) Ltd, August 1996.

D3586 Crewe 1958 15A 9/85 P D3586

08471 to Severn Valley Railway, Bridgnorth; despatched from BR Swindon Works, 14th April 1986.

D3587 Crewe 1958 ? ? P 08472

08472 to Great North Eastern Railway and allocated to Bounds Green Depot, London; sold to RFS (Engineering) Ltd, Doncaster, 1998, but with the locomotive remaining at Bounds Green Depot, London, on hire; to GNER, Craigentinny Depot, Edinburgh, on hire, 1999; to Wabtec, Doncaster, December 2000; to GNER, Craigentinny Depot, Edinburgh, on hire, by August 2003; returned to Wabtec, Doncaster, by 1st April 2004; to GNER, Craigentinny Depot, Edinburgh, on hire, 10th August 2005; to Wabtec, Doncaster, 26th November 2010; to Craigentinny Depot, Edinburgh, on hire, 12th September 2011.

D3588 Crewe 1958 86A 3/86 P 08473

08473 to T.J. Thomson Ltd, Stockton-on-Tees; despatched from Leicester, 28th November 2000; partly scrapped; remains sold to Dean Forest Railway, Lydney, for spares, 5th March 2001.

D3591 Crewe 1958 62A 9/85 P D3591

08476 to Swanage Railway, Dorset; despatched from BR Swindon Works, 21st March 1986; to Battlefield Line, Shackerstone, 24th July 2009; to Swanage Railway, Dorset, 10th March 2012.

D3594 Horwich 1958 86A 11/91 P 08479 / 13594

08479 to East Lancashire Railway, Bury, 30th April 1993.

D3595 Horwich 1958 16A ? P 08480

08480 to Railway Support Services, Wishaw, Warwickshire; despatched from DB

Cargo, Toton Depot, 1st February 2017; to Great Central Railway, Loughborough, for gala held on 18th and 19th March 2017; to Crown Point Depot, Norwich, on hire, 18th July 2017.

D3596 Horwich 1958 86A 4/02 F 08481
08481 to Barry Rail Centre, Barry Island; despatched from Wigan, 7th October 2005; to European Metal Recycling, Kingsbury, for scrap, May 2011; scrapped early June 2011.

D3599 Horwich 1958 ZN 6/95 P 08484
08484 to Railcare Ltd, Wolverton Works, June 1995; locomotive included in sale when BRML works privatised; to LH Group, Barton under Needwood, for repair, 16th November 2003; returned to Wolverton Works, 10th February 2004; sold to Traditional Traction, Wishaw, Warwickshire, April 2006; to Port of Felixstowe, on hire, 25th April 2006; to Traditional Traction, Wishaw, for storage, about 10th November 2006; to Nene Valley Railway, Wansford, for gala, February 2007; returned to Traditional Traction, February 2007; to St Philip's Marsh Depot, Bristol, for tyre turning, 19th April 2007; to Port of Felixstowe, on hire, week-ending 4th May 2007; to Tyseley Steam Depot, Birmingham, for tyre turning, 13th July 2012; to Gloucestershire Warwickshire Railway, Toddington, on hire, 14th July 2012; to Port of Felixstowe, on hire, 8th November 2013; to Traditional Traction, Wishaw, 9th September 2014; to National Railway Museum, Shildon, on hire, 16th September 2015; to Network Rail, Old Dalby Test Centre, on hire, 25th February 2016; to Bombardier Transportation, Litchurch Lane Works, Derby, on hire, 5th August 2016; to Railway Support Services, Wishaw, 17th October 2016; to Cemex Rail Products, Washwood Heath, on hire, 25th October 2016; to Hitachi, Newton Aycliffe, on hire, 4th April 2018.

D3600 Horwich 1958 12A 11/07 P 08485
08485 to West Coast Railway Company, Carnforth, 1st September 2010.

D3605 Horwich 1958 62A 11/85 P D3605
08490 to Strathspey Railway, Aviemore; despatched from BR Perth, 18th June 1987.

D3607 Horwich 1958 66B ? F 08492
08492 to Barrow Hill Engine Shed Society (HNRC), Staveley; despatched from EWS Motherwell Depot, 2nd June 2006; arrived BHESS, 5th June 2006; to European Metal Recycling, Kingsbury, for scrap, 14th February 2012; scrapped, 15th February 2012.

D3608 Doncaster 1958 86A 7/99 F 08493
08493 sold to RT Rail of Crewe; despatched from Canton Depot, Cardiff, and moved direct to Wabtec, Doncaster, for storage, September 2003; to C.F. Booth Ltd, Rotherham, for scrap, 19th May 2008; scrapped, August 2008.

D3610 Doncaster 1958 5A ? P 08495
08495 to North Yorkshire Moors Railway, Grosmont; despatched from DB Cargo, Crewe Electric Depot, 22nd December 2016.

D3613 Darlington 1958 40A 2/69 F DAVID
to NCB Bestwood Colliery, Nottinghamshire, August 1969; to Linby Colliery, by September 1971; to Moor Green Colliery, Newthorpe, November 1971; to 16A Toton Depot, for

repairs, by 26th October 1975; returned to Moor Green Colliery, Newthorpe, by 22nd December 1975; scrapped on site by The Vic Berry Company of Leicester, 10th April 1985.

D3618 Darlington 1958 40A 4/69 F ROBIN / D16
to NCB Bestwood Colliery, Nottinghamshire, August 1969; to Annesley Colliery, March 1970; to BR Toton Depot for repairs, June 1974; returned to Annesley Colliery; to Cotgrave Colliery, 24th July 1980; to Moor Green Colliery, Newthorpe, 30th March 1981; scrapped on site by The Vic Berry Company of Leicester, 26th March 1985.

D3619 Darlington 1958 40A 2/69 F SIMON / D15
to NCB Gedling Colliery, Nottinghamshire, September 1969; to Bestwood Colliery, November 1969; to Linby Colliery, about July 1971; to Moor Green Colliery, Newthorpe, 24th November 1975; scrapped on site by The Vic Berry Company of Leicester, week-ending 5th April 1985.

D3638 Darlington 1958 52A 11/70 F 9185-61
to NCB Bates Colliery, Blyth, on hire, 19th November 1970; sold to NCB, 28th January 1971; to BR Gateshead Depot, for repairs, 4th February 1971; to BR Cambois Depot, Blyth, for repairs, 12th February 1971; to Bates Colliery, Blyth, 16th February 1971; to Ashington Colliery, 15th April 1971; to Bates Colliery, Blyth, 22nd April 1971; to Ashington Central Workshops, March 1975; used for spares; remains scrapped, September 1975.

D3642 Darlington 1958 36C 6/69 F 37
to BSC Redbourn Works, Scunthorpe, September 1969; to BSC Appleby-Frodingham Works, Scunthorpe, October 1975; scrapped, October 1978.

D3648 Darlington 1959 52A 1/71 F 9185-60
to NCB Bates Colliery, Blyth, on hire, 16th February 1971; to BR Cambois Depot, Blyth, for repairs, 27th February 1971; to Bates Colliery, Blyth, 1st March 1971; sold to NCB, 1st March 1971; to BR Cambois Depot, Blyth, for further repairs, 5th March 1971; to Bates Colliery, Blyth, 23rd March 1971; scrapped on site by L. Marley & Co Ltd of Stanley, February 1977.

D3654 Doncaster 1958 81A 6/04 P 08499 / REDLIGHT
08499 to Pullman Rail, Canton Depot, Cardiff (locomotive included in sale when depot sold), June 2005.

D3655 Doncaster 1958 ? ? P 08500
08500 sold to Harry Needle Railroad Company; to Nemesis Rail, Burton upon Trent; despatched from DB Schenker, Tees Yard, 23rd October 2015.

D3657 Doncaster 1958 51L 10/88 P 08502/
08502 LYBERT DICKINSON
to ICI Wilton Works, Middlesbrough, 6th September 1988; sold to Harry Needle Railroad Company, February 2007; to Barrow Hill Engine Shed Society, Staveley, for repaint, 3rd August 2007; to Northern Rail Ltd, Heaton Depot, Newcastle upon Tyne, on hire, 10th September 2007; to Barrow Hill Engine Shed Society, Staveley, 21st June 2012; to Flixborough Wharf Ltd, Flixborough, Scunthorpe, on hire, October 2013; to Barrow Hill

Engine Shed Society, Staveley, for repairs, 26th May 2014; to Tyseley Steam Depot, Birmingham, for tyre turning, 3rd November 2015; to Moveright International, Wishaw, Warwickshire, for storage, 20th December 2015; to GBRf, Speke, Liverpool, on hire, 20th December 2015; to Barrow Hill Engine Shed Society, Staveley, 20th July 2017; to GBRf, c/o Ford Motor Co Ltd, Speke, Liverpool, on hire, late 2017; to Barrow Hill Engine Shed Society, Staveley, 10th April 2019; to East Kent Railway, Shepherdswell, 24th April 2019.

D3658 Doncaster 1958 55A 10/88 P 08503

08503 to ICI Wilton Works, Middlesbrough; despatched from BR Doncaster Works, 26th September 1988; sold to Harry Needle Railroad Company, February 2007; to Heanor Heavy Haulage, Langley Mill, for storage, 7th December 2007; to Moveright International, Wishaw, Warwickshire, for storage, late 2008; sold to Traditional Traction, April 2013; to Colne Valley Railway, Castle Hedingham, 15th April 2013; to Barry Rail Centre, Barry Island, on hire, 22nd October 2013; to St Philip's Marsh Depot, Bristol, for tyre turning, 11th June 2019; returned to Barry Rail Centre, 12th June 2019.

D3662 Doncaster 1958 81D 8/93 P 08507

08507 to S.M. McGregor & Sons, Bicester, for store, about April 1994; to South Yorkshire Railway Preservation Society (HNRC), Meadowhall, Sheffield, 23rd June 1995; to Barrow Hill Engine Shed Society, Staveley, 3rd November 1999; to Bombardier Transportation, Central Rivers Depot, Barton under Needwood, on hire, 7th April 2001; to Barrow Hill Engine Shed Society, Staveley, for repairs, 10th January 2011; to Port of Boston, on hire, 13th June 2011; sold to Riviera Trains Ltd of Crewe; to Crewe Diesel Depot, 19th September 2013; to Nemesis Rail, Burton upon Trent, 12th January 2018.

D3673 Darlington 1958 66B 3/04 P 08511

08511 to Traditional Traction, Wishaw, Warwickshire; despatched from EWS Ayr Depot, 29th June 2007; to Barry Rail Centre, Barry Island, 9th July 2007; to Port of Felixstowe, on hire, 8th July 2009; to Wembley Depot, London, for tyre turning, 2nd July 2010; to Colne Valley Railway, Castle Hedingham, for painting, 2nd August 2010; to Port of Felixstowe, on hire, 12th January 2011; to Tilbury Docks, on hire, 8th November 2013; to Traditional Traction, Wishaw, 2014; to Mid-Norfolk Railway, Dereham, 1st September 2014; to Port of Felixstowe, on hire, 8th September 2014; to Whatley Quarry, Somerset, on hire, 11th May 2015; to Moveright International, Wishaw, for repairs, 3rd July 2015; to Port of Felixstowe, on hire, 7th July 2015; to Moveright International, Wishaw, for repairs, December 2015; to Port of Boston, on hire, 7th February 2016; to Old Dalby Test Centre, on hire, 17th February 2016; to National Railway Museum, Shildon, 25th February 2016; to Moveright International, Wishaw, for repairs, 25th March 2016; to Bombardier Transportation, Litchurch Lane Works, Derby, on hire, 4th April 2016; to Traditional Traction, Wishaw, August 2016; to Chasewater Railway, Staffordshire, 12th September 2016; to Railway Support Services, Wishaw, 28th January 2017; to Arriva Traincare, Cambridge Depot, on hire, 23rd February 2017; to Arriva Traincare, Eastleigh, on hire, 10th December 2018; to GBRf, East Yard, Eastleigh, on hire, 28th March 2019.

D3677 Darlington 1958 52A 1/92 F 08515

08515 to Gwent Demolition, 1995; firm went bankrupt and sale cancelled; resold to T.J. Thomson, Stockton-on-Tees, for scrap, April 2000; dismantled at Gateshead Depot, February 2001; remains to T.J. Thomson, Stockton-on-Tees, February 2001; to Foster

Yeoman Quarries Ltd, Merehead Stone Terminal; used for spares, 2001; remains scrapped, February 2001.

D3678 Darlington 1958 16A 2/04 P 08516

08516 to Arriva Traincare, Barton Hill Depot, Bristol (ex DB Schenker, with site), April 2011.

D3679 Darlington 1958 30A 9/93 F 08517

08517 to Barrow Hill Engine Shed Society (HNRC), Staveley; despatched from Stratford Depot, 4th June 2001; to Wabtec, Doncaster, 27th June 2001; rebuilt using spares from 08668; stored in West Yard, Doncaster, late 2002; to C.F. Booth Ltd, Rotherham, 2007, but not scrapped: to Harry Needle Railroad Company, Long Marston, 19th December 2007; used for spares, early 2011; to C.F. Booth Ltd, Rotherham, for scrap, 1st July 2011; scrapped, 21st July 2011.

D3685 Doncaster 1958 ? ? P 08523 / HO61

08523 to RT Rail, Crewe, and stored at Crewe Diesel Depot, 2004; to L&NWR, Carriage Works, Crewe, for overhaul, 4th March 2004; to Heritage Centre, Crewe, by July 2005; to Hays Chemicals, Sandbach, on hire, about July 2005; to RMS Locotec, Wakefield, for repairs, February 2007; to Celtic Energy, Onllwyn Disposal Point, on hire, 30th March 2007; RT Rail acquired by RMS Locotec, 8th November 2007; to RMS Locotec, Wakefield, November 2007; to Celtic Energy, Onllwyn Disposal Point, on hire, by 17th November 2007; to RMS Locotec, Weardale Railway, Wolsingham, for repairs, 11th January 2012; to PD Ports, Tees Dock, on hire, 22nd August 2012; to RMS Locotec, Weardale Railway, Wolsingham, for repairs, 4th December 2012; to Scot Rail, Inverness, on hire, 28th August 2013.

D3689 Darlington 1959 ZI 6/95 P 08527

08527 to ABB (Customer Support) Ltd, Ilford, June 1995; locomotive included in sale when BRML works privatised; sold to Harry Needle Railroad Company, 2006; to Barrow Hill Engine Shed Society (HNRC), Staveley, September 2006; to Roberts Road Depot, Doncaster, on hire, 7th June 2007; to Flixborough Wharf Ltd, Flixborough, Scunthorpe, on hire, 31st August 2010; to Barrow Hill Engine Shed Society, Staveley, October 2013; to GBRf, Trafford Park, Manchester, on hire, 10th February 2014; to Northern Rail, Allerton Depot, Liverpool, on hire, 28th May 2014; to Barrow Hill Engine Shed Society, Staveley, 11th August 2016; to GBRf, Immingham, on hire, December 2016; to Attero Recycling Ltd, Rossington, Doncaster, on hire, 19th January 2019.

D3690 Darlington 1959 2F 7/05 P D3690

08528 to Battlefield Line, Shackerstone, 31st July 2010; to Great Central Railway, Loughborough, 3rd July 2014.

D3691 Darlington 1959 40B ? F 08529

08529 to RT Rail, Crewe; despatched from Doncaster Depot, April 2005; to Wabtec, Doncaster, for repairs, July 2005; to RMS Locotec, Dewsbury, August 2005; to Wabtec, Doncaster, for storage, by November 2005; to C.F. Booth Ltd, Rotherham, for scrap, 21st May 2008; scrapped, August 2008.

D3692 Darlington 1959 ? ? P 08530

08530 to Traditional Traction, Wishaw, Warwickshire, about August 2006; to Port of Felixstowe, on hire, about January 2007; to LH Group, Barton under Needwood, for repairs, 24th December 2007; sold to Freightliner, March 2009, and based initially at Southampton Docks.

D3699 Darlington 1959 2F 10/00 F 08535

08535 to RT Rail, Crewe; despatched from Crewe Diesel Depot and stored at L&NWR, Carriage Works, Crewe, 4th March 2004; to RMS Locotec, Wakefield, for repairs, 5th September 2007; sold to RMS Locotec, November 2007; to Corus, Shotton Steelworks, 21st January 2009; used for spares; remains to C.F. Booth Ltd, Rotherham, for scrap, 15th September 2009; scrapped, 17th September 2009.

D3700 Darlington 1959 16C 6/95 P 08536

08536 to Railway Vehicle Engineering Ltd, Railway Technical Centre, Derby; despatched from East Midlands Trains, Etches Park, Derby, July 2010; sold to Railway Support Services and moved to Wishaw, Warwickshire, 9th October 2018.

D3723 Darlington 1959 1A 7/90 P 08556

08556 to North Yorkshire Moors Railway, Grosmont, 23rd October 1993.

D3734 Crewe 1959 5A ? P 08567

08567 to Arlington Fleet Services, Eastleigh Works, Hampshire; despatched from DB Schenker, Crewe Electric Depot, 3rd December 2015.

D3735 Crewe 1959 ZH 6/95 P 08568 / ST ROLLOX

08568 to Railcare Ltd, Springburn Works, Glasgow, June 1995; locomotive included in sale when BRML works privatised; sold to Railway Support Services, May 2017; to Railway Support Services, Wishaw, Warwickshire, 8th December 2017; to Tyseley Steam Depot, Birmingham, for tyre turning, 6th June 2019; returned to Railway Support Services, Wishaw, June 2019.

D3738 Crewe 1959 ? ? P 08571

08571 to Great North Eastern Railway and allocated to Craigentinny Depot, Edinburgh; sold to RFS (Engineering) Ltd, Doncaster, about March 1997; to GNER, Craigentinny Depot, Edinburgh, on hire, about April 1997; returned to RFS (Engineering) Ltd, 1998; to ARC, Whatley Quarry, on hire, January 1999; returned to RFS (Engineering) Ltd, Doncaster, by 17th February 1999; to Hanson, Whatley Quarry, on hire, by 19th September 1999; to Merehead Stone Terminal, September 2001; to Wabtec, Doncaster, by 26th January 2002; to GNER, Bounds Green Depot, London, on hire, about March 2005; to LH Group, Barton under Needwood, for repairs, 24th August 2015; to Bounds Green Depot, London, on hire, 17th October 2015; to Wabtec, Doncaster, for repairs, 3rd February 2016; to Bombardier Transportation, Litchurch Lane Works, Derby, on hire, March 2016; to LH Group, Barton under Needwood, 1st August 2016; to Freightliner, Port of Felixstowe, on hire, 26th August 2016; to Wabtec, Doncaster, for repairs, 29th August 2016; to LH Group, Barton under Needwood, 29th September 2016; to Freightliner, Port of Felixstowe, on hire, 9th November 2016; returned to LH Group, Barton under Needwood; to Freightliner, Southampton Docks, on hire, 8th May 2017; to LH Group,

Barton under Needwood, 14th August 2017; to Daventry International Rail Freight Terminal, on hire, 19th October 2017.

D3740 Crewe 1959 ZI 6/95 P 08573
08573 to ABB (Customer Support) Ltd, Ilford, June 1995; locomotive included in sale when BRML works privatised; sold to RT Rail, Crewe, about January 2001; to Wabtec, Doncaster, for repairs, 25th October 2001; to Channel Tunnel Rail Link, Beechbrook Farm, near Ashford, on hire, 9th November 2001; to Wabtec, Doncaster, for repairs, by 10th January 2003; to Bombardier Transportation, Ilford, on hire, by September 2003; to Wabtec, Doncaster, for repairs, August 2004; to Freightliner, Coatbridge, on hire, 12th March 2005; to Scot Rail, Inverness, on hire, April 2005; to Wabtec, Doncaster, 8th May 2006; to Tubelines, Ruislip, London, on hire, 17th May 2006; sold to RMS Locotec, Wakefield, November 2006; to Wabtec, Doncaster, for repairs, about May 2007; to Bombardier Transportation, Ilford Depot, on hire, 12th July 2007; to RMS Locotec, Weardale Railway, Wolsingham, 6th March 2018.

D3743 Crewe 1959 86A 6/00 F 08576
08576 to Battlefield Line, Shackerstone; despatched from Canton Depot, Cardiff, 12th February 2004; used for spares; remains to T.J. Thomson Ltd, Stockton-on-Tees, for scrap, 8th June 2006; scrapped, May 2007.

D3745 Crewe 1959 16A ? P 08578
08578 to Harry Needle Railroad Company, Long Marston; despatched from DB Schenker, Toton Depot, 18th January 2016.

D3747 Crewe 1959 2F ? P 08580
08580 to Railway Support Services, Wishaw, Warwickshire; despatched from Bescot Depot, September 2015; to Colne Valley Railway, Castle Hedingham, 25th February 2016; to Railway Support Services, Wishaw, 25th May 2017; to Castle Cement Works, Ketton, for open weekend, 29th June 2018; to Bounds Green Depot, London, on hire, 4th July 2018.

D3755 Crewe 1959 55H ? P 08588 / HO47
08588 to RMS Locotec, Dewsbury; despatched from Neville Hill Depot direct to Wabtec, Doncaster, for repairs, 16th April 2005; sold to Wabtec, Doncaster, May 2005; to Network Rail, Whitemoor Yard, March, on hire, 15th September 2005; sold to RT Rail, Crewe, about March 2006; to Wabtec, Doncaster, for repairs, 31st March 2006; to RMC Aggregates, Dove Holes, on hire, 7th April 2006; to Channel Tunnel Rail Link, Dagenham, London, on hire, 28th September 2006; to Wabtec, Doncaster, for repairs, 6th November 2006; to Bombardier Transportation, Ilford Depot, on hire, 20th November 2006; sold to RMS Locotec, Wakefield, November 2007; to Network Rail, Whitemoor Yard, March, on hire, December 2007; to PD Ports, Tees Dock, on hire, July 2008; to RMS Locotec, Weardale Railway, Wolsingham, 1st March 2011; to Cemex Rail Products, Washwood Heath, on hire, 10th April 2012; to Tyseley Steam Depot, Birmingham, for tyre turning, 18th January 2017; to RMS Locotec, Weardale Railway, Wolsingham, 20th January 2017; to Heaton Depot, Newcastle upon Tyne, on hire, 16th March 2017; to RMS Locotec, Weardale Railway, Wolsingham, 11th October 2017; to Loram UK Ltd, Railway Technical Centre, Derby, on hire, 13th February 2018.

D3757 Crewe 1959 52B 9/93 P 08590 / RED LION

08590 to Midland Railway, Butterley, Derbyshire, July 1994.

D3759 Crewe 1959 ? 3/15 P 08993 / ASHBURNHAM

08993 cut down cab locomotive; formerly numbered 08592 until 13th April 1985; to Keighley & Worth Valley Railway, Haworth; despatched from DB Schenker, Stoke on Trent, 1st September 2015.

D3760 Crewe 1959 5A ? P 08593

08593 to Railway Support Services, Wishaw, Warwickshire; despatched from DB Schenker, Crewe Electric Depot, 20th January 2016; used for spares.

D3761 Crewe 1959 16A 2/97 F 08594

08594 to Mike Darnall, Newton Heath, Manchester, November 2000; to Wabtec, Doncaster, by April 2001; to Mike Darnall, 9th January 2008; to Traditional Traction, Wishaw, Warwickshire, 15th July 2010; to European Metal Recycling, Kingsbury, for scrap, 21st July 2010; scrapped, September 2010.

D3763 Derby 1959 81A 3/77 P 08596

08596 to Bowaters UK Paper Co Ltd, Sittingbourne, Kent; despatched from 81A Old Oak Common Depot, 16th May 1977; to BR Swindon Works, for overhaul, 20th November 1981; returned to Bowaters UK Paper Co Ltd, 4th January 1982; sold to RFS (Engineering) Ltd, Kilnhurst, 6th June 1991; to ECC, Quidhampton, on hire, 29th July 1991; returned to RFS (Engineering) Ltd, Doncaster, 5th to 7th September 1991; to Channel Tunnel, Folkestone, (Balfour Beatty Ltd), on hire, 21st September 1991; to RFS (Engineering) Ltd, Doncaster, 6th July 1993; to Sheerness Steel Co Ltd, Sheerness, Kent, on hire, July 1993; returned to RFS (Engineering) Ltd, Doncaster, by 1st November 1996; to EWS, Decoy Yard, Doncaster, on hire, May 1999; returned to RFS (Engineering) Ltd, Doncaster; to Balfour Beatty, Leeds Station contract, on hire, about March 2000; returned to RFS (Engineering) Ltd, Doncaster, by 14th May 2001; to Blue Circle, Hope Cement Works, Derbyshire, on hire, about February 2002; returned to Wabtec, Doncaster, November 2002; to GNER, Bounds Green Depot, London, on hire, by May 2003; to Wabtec, Doncaster, by 27th July 2003; to Mendip Rail, Whatley Quarry, on hire, 5th January 2004; returned to Wabtec, Doncaster, by 7th March 2004; to Daventry International Rail Freight Terminal, on hire, by 5th October 2004; to Wabtec, Doncaster, for repairs, 12th November 2004; to Daventry International Rail Freight Terminal, on hire, by 8th December 2004; to Wabtec, Doncaster, for repairs, by 12th March 2005; to Hanson Quarry Products, Whatley Quarry, on hire, by 22nd January 2006; to Wabtec, Doncaster, by 5th February 2006; to Bounds Green Depot, London, on hire, 16th February 2008; to LH Group, Barton under Needwood, for repairs, 5th December 2014; to Bounds Green Depot, London, on hire, 1st July 2015; to Railway Support Services, Wishaw, Warwickshire, 4th February 2016; to Craigentinny Depot, Edinburgh, on hire, 11th February 2016.

D3765 Derby 1959 5A 11/86 P 08598 / HO16

08598 to Powell Duffryn Fuels Ltd, NCBOE Gwaun-cae-Gurwen, 16th January 1987; sold to RMS Locotec, Dewsbury, September 1995; remained on hire to Celtic Energy, Gwaun-cae-Gurwen; to RMS Locotec, Dewsbury, September 1998; to Cleveland Potash

Ltd, Boulby Mine, on hire, 21st June 1999; returned to RMS Locotec, Dewsbury, by 14th July 2000; sold to Potter Group, Knowsley, September 2001; to Potter Group, Selby, 24th May 2002; to Potter Group, Knowsley, by 12th April 2003; to Gloucestershire Warwickshire Railway, Toddington, for gala, 7th July 2010; to Potter Group, Ely, week-ending 3rd September 2010; acquired by Ed Murray & Sons Ltd, Hartlepool, mid-2013; to Chasewater Railway, Staffordshire, 10th September 2013; purchased by A.V. Dawson Ltd; to A.V. Dawson Ltd, Middlesbrough, 4th May 2017; to Railway Support Services, Wishaw, for overhaul, 19th July 2019.

D3767 Derby 1959 30A 11/97 P 08600
08600 to A.V. Dawson Ltd, Middlesbrough; despatched from Eastleigh, 17th November 1997; to EWS Thornaby Depot, for tyre turning, 14th June 2004; returned to A.V. Dawson Ltd, Middlesbrough, 1st July 2004; to LH Group, Barton under Needwood, for repairs, 14th October 2008; returned to A.V. Dawson Ltd, Middlesbrough, 23rd March 2009.

D3769 Derby 1959 12A 3/86 P 004
08602 to RFS (Engineering) Ltd, Kilnhurst, 30th June 1988; to Foster Yeoman Quarries Ltd, Isle of Grain Stone Terminal, on hire, December 1988; to Sheerness Steel Co Ltd, Sheerness, Kent, on hire, December 1989; to ABB Transportation, Litchurch Lane Works, Derby, on hire, March 1990; to ABB Transportation, York, on hire, November 1990; to ABB Transportation, Litchurch Lane Works, Derby, on hire, November 1990; sold to ABB Transportation, Litchurch Lane Works, Derby, April 1991; to Fragonset, Derby, for overhaul, about September 2003; to Bombardier Transportation, Litchurch Lane Works, Derby, about April 2004.

D3771 Derby 1959 16C 7/93 P 604 / PHANTOM
08604 to Didcot Railway Centre, Oxfordshire, 28th September 1994.

D3772 Derby 1959 8F ? P G.R. WALKER
08605 sold to Riviera Trains Ltd of Crewe; despatched from Springs Branch Depot, Wigan, early 2017; to Ecclesbourne Valley Railway, Wirksworth, on loan, 7th June 2018.

D3778 Derby 1959 ? ? P 08611
08611 acquired by Alstom Transport, Edge Hill Depot, Liverpool, 2007; to Alstom Transport, Longsight Depot, Manchester, by 27th March 2011; to Arlington Fleet Services, Eastleigh Works, for overhaul, by 13th January 2016; to Alstom Transport, Stonebridge Park Carriage Maintenance Depot, London, 20th July 2016.

D3780 Derby 1959 8J 12/93 P 08613 / HO64
08613 to Trafford Park Estates Ltd, Manchester, 2nd February 1994; sold to Wabtec, Doncaster, November 2000; overhauled; to Trafford Park, Manchester, on hire, 18th December 2000; sold to RT Rail, Crewe, January 2001; to Bombardier Transportation, Ilford Depot, on hire, March 2001; to Wabtec, Doncaster, for overhaul, November 2006; returned to Ilford Depot, on hire; sold to RMS Locotec, 2007; to RMS Locotec, Wakefield, for repairs, 8th May 2007; to Barrow Hill Engine Shed Society, Staveley, 5th October 2007; to Corus, Shotton, on hire, November 2007; to Castle Cement Works, Ketton, on hire, 8th April 2009; to Boden Rail Engineering, Washwood Heath, for repairs, November 2011; to Celtic Energy, Onllwyn Disposal Point, on hire, 11th January 2012; to RMS

Locotec, Weardale Railway, Wolsingham, 4th January 2017; to PD Ports, Tees Dock, on hire, 3rd October 2018; to RMS Locotec, Weardale Railway, Wolsingham, for repairs, 11th July 2019.

D3782 Derby 1959 8J 12/93 P 08615 / UNCLE DAI
08615 to Trafford Park Estates Ltd, Manchester, 2nd February 1994; sold to Wabtec, Doncaster, November 2000; to Hanson Quarry Products, Whatley Quarry, on hire, by 5th May 2002; to Merehead Stone Terminal, on hire, October 2002; returned to Wabtec, Doncaster, 12th August 2003; to GNER, Craigentinny Depot, Edinburgh, on hire, by March 2004; to LH Group, Barton under Needwood, 11th February 2016; to Tata Steel, Shotton, on hire, 31st January 2019.

D3784 Derby 1959 ? ? P 08617
08617 acquired by Alstom Transport, Wembley Depot, London, 2007; to Arlington Fleet Services, Eastleigh Works, for overhaul, 29th March 2016; to Alstom Transport, Oxley Depot, Wolverhampton, about 27th September 2017.

D3785 Derby 1959 52A 9/90 F 08618
08618 to Gwent Demolition, 1995; firm went bankrupt and sale cancelled; resold to T.J. Thomson Ltd, Stockton-on-Tees, for scrap, April 2000; dismantled at Gateshead Depot, for spares, April 2001; to T.J. Thomson Ltd, Stockton-on-Tees, chassis only, April 2001; to Freightliner, Southampton Docks, for spares, April 2001; scrapped on site (by Southampton Steel Ltd), September 2003.

D3789 Derby 1959 66B 12/95 P 08622 / HO28 / 19
08622 to RMS Locotec, Dewsbury, 4th May 2002; to Flixborough Wharf Ltd, Flixborough, Scunthorpe, on hire, 6th March 2003; to RMS Locotec, Dewsbury, 19th April 2004; to PD Ports, Tees Dock, on hire, 6th August 2004; to RMS Locotec, Dewsbury, 14th February 2006; to RMS Locotec, Wakefield, 29th June 2006; to Flixborough Wharf Ltd, Flixborough, Scunthorpe, on hire, 9th October 2006; to Corus, Trostre Works, Llanelli, on hire, 3rd January 2008; to PD Ports, Tees Dock, on hire, week commencing 9th November 2009; to RMS Locotec, Weardale Railway, Wolsingham, for repairs, 18th January 2011; to Castle Cement Works, Ketton, on hire, 1st March 2011; to Nene Valley Railway, Wansford, for gala, September 2011; to Castle Cement Works, Ketton, on hire, October 2011; to RMS Locotec, Weardale Railway, Wolsingham, 20th November 2012; to Castle Cement Works, Ketton, on hire, 31st January 2013; to RMS Locotec, Weardale Railway, Wolsingham, for repairs, 7th November 2016; to Castle Cement Works, Ketton, on hire, 6th July 2018.

D3790 Derby 1959 2F ? P 08623
08623 to Harry Needle Railroad Company; despatched from Bescot Depot; to Hope Cement Works, Derbyshire, on hire, 31st August 2017.

D3792 Derby 1959 2F 1/98 F 08625
08625 to Dean Forest Railway, Lydney; despatched from Canton Depot, Cardiff, 17th June 2000; to Cotswold Rail, Moreton in Marsh, about June 2001; to European Metal Recycling, Kingsbury, for scrap, February 2004; scrapped, 26th February 2004.

D3795 Derby 1959 2F 8/99 F 08628

08628 to Goodman, Wishaw, Warwickshire; despatched from Saltley Depot, 28th September 2005; to European Metal Recycling, Kingsbury, about October 2005; to Goodman, Wishaw, 13th April 2006; to Bryn Engineering, and stored at Redrock Plant & Truck Services, Blackrod, Bolton, April 2006; to Ribble Steam Railway, Preston, 5th August 2011; to Moveright International, Wishaw, Warwickshire, 19th November 2015; used for spares; remains to European Metal Recycling, Kingsbury, for scrap, 19th May 2016; scrapped.

D3796 Derby 1959 ZN 6/95 P 08629 / WOLVERTON

08629 to Railcare Ltd, Wolverton Works, June 1995; locomotive included in sale when BRML works privatised; to Great Central Railway, for gala, 11th February 2011; to Alstom Transport, Wolverton Works, 14th February 2011; to Chinnor & Princes Risborough Railway, Oxfordshire, for gala, 1st October 2013; to Knorr Bremse Rail Services Ltd, Wolverton Works, 8th October 2013.

D3797 Derby 1959 ? ? P CELSA 3

08630 to Barrow Hill Engine Shed Society (HNRC), Staveley, 17th February 2016; overhauled; to Celsa, Cardiff, on hire, 23rd August 2016; left Celsa, 19th April 2018; stored in haulier's yard; to Traditional Traction, Wishaw, Warwickshire, for repairs, 26th April 2018; to Celsa, Cardiff, on hire, 11th June 2018; to Barrow Hill Engine Shed Society, Staveley, for repairs, 25th February 2019; to Celsa, Cardiff, on hire, week-ending 23rd March 2019.

D3798 Derby 1959 31B 12/92 P 08631 / EAGLE

08631 to Great Eastern Railway Company, County School Station, Norfolk, spring 1994; to Great Eastern Traction, Hardingham, Norfolk, 12th August 1995; to Fragonset, Derby, 13th December 1997; to Bombardier Transportation, Litchurch Lane Works, Derby, on hire, December 2003; returned to Fragonset (FM Rail), Derby, about June 2004; to Mid-Norfolk Railway, Dereham, on hire, 22nd March 2007; returned to FM Rail, Derby; to Gwili Railway, Bronwydd Arms, on hire, 25th May 2010; to Mid-Norfolk Railway, Dereham, 29th June 2011; to Nemesis Rail, Burton upon Trent, 27th May 2014; to Bombardier Transportation, Litchurch Lane Works, Derby, 2nd June 2014; to Nemesis Rail, Burton upon Trent, 22nd April 2015; sold to Locomotive Storage Ltd and moved to Crewe Diesel Depot, 1st June 2015; to RMS Locotec, Wolsingham, for repairs, 27th February 2019.

D3799 Derby 1959 ? ? P 08632

08632 to Railway Support Services, Wishaw, Warwickshire; despatched from DB Cargo, Mossend, January 2017; to Knorr Bremse Rail Services Ltd, Springburn Works, Glasgow, on hire, 19th January 2017; to Chasewater Railway, Staffordshire, 21st February 2017; to Railway Support Services, Wishaw, for repairs, March 2017; to Tata Steel, Trostre Works, Llanelli, on hire, 24th May 2017; to Railway Support Services, Wishaw, 28th July 2017; to East Midlands Trains, Neville Hill Depot, Leeds, on hire, August 2017; to Railway Support Services, Wishaw, 22nd May 2019; to GBRf, East Yard, Eastleigh, on hire, 3rd June 2019.

D3800 Derby 1959 ? ? P D3800

08633 to Churnet Valley Railway, Cheddleton; despatched from DB Wagon Works, Stoke on Trent, 6th December 2016.

D3801 Derby 1959 30A 2/93 F 08634

08634 to Barrow Hill Engine Shed Society (HNRC), Staveley, 2nd July 2001; to West Coast Railway Company, Carnforth, 11th July 2002; scrapped, February 2005.

D3802 Derby 1959 81A 2/04 P 08635

08635 to T.J. Thomson, Stockton-on-Tees; despatched from EWS Toton Depot, 28th February 2007; to Severn Valley Railway, Bridgnorth, 26th April 2007.

D3807 Horwich 1958 ? ? P 08640

08640 to Traditional Traction, Wishaw, Warwickshire, December 2016; to Axiom Rail, Wagon Works, Stoke on Trent, on hire, 4th January 2017.

D3810 Horwich 1959 82B ? P 08643

08643 to Foster Yeoman Quarries Ltd, Merehead Stone Terminal, 16th April 2003; to Whatley Quarry, by 15th April 2004; to Merehead Stone Terminal, by June 2005; to Isle of Grain Stone Terminal, Kent, by February 2007; to West Somerset Railway, Minehead, for gala, 12th June 2008; to Merehead Stone Terminal, 14th June 2008; to Whatley Quarry, by 11th June 2009; to Merehead Stone Terminal, by 9th June 2011.

D3814 Horwich 1959 36A 4/97 F 08647

08647 to South Yorkshire Railway Preservation Society, Meadowhall, Sheffield; despatched from Adtranz, Crewe, 12th March 1997; used for spares, 7/8th November 1998; remains to Mayer Parry Ltd, Snailwell, Cambridgeshire, for scrap, 20th November 1998; scrapped.

D3815 Horwich 1959 84A ? P 08648 / HO65

08648 purchased by RT Rail, Crewe; despatched from Laira Depot, Plymouth, and moved direct to Wabtec, Doncaster, for repairs, 24th July 2002; to Midland Road Depot, Leeds, for fitting with auto couplers, mid-2004; to Brunner Mond, Winnington Works, Northwich, on hire, about September 2004; to Wabtec, Doncaster, for repairs, about September 2005; to Brunner Mond, Northwich, on hire, 13th January 2006; to Wabtec, Doncaster, for repairs, August 2007; sold to RMS Locotec, Wakefield, November 2007; to PD Ports, Tees Dock, on hire, 17th January 2011; to RMS Locotec, Weardale Railway, Wolsingham, 27th August 2014; to Tyseley Steam Depot, Birmingham, for tyre turning, 24th November 2014; to Grand Central Railway, Heaton Depot, Newcastle upon Tyne, on hire, 27th November 2014; to RMS Locotec, Weardale Railway, Wolsingham, for repairs, 17th March 2017; to Grand Central Railway, Heaton Depot, Newcastle upon Tyne, on hire, 11th October 2017; to RMS Locotec, Weardale Railway, Wolsingham, 7th March 2018; to Scot Rail, Inverness, on hire, 11th April 2018.

D3816 Horwich 1959 ZG 6/95 P 08649 / BRADWELL

08649 to Wessex Traincare Ltd, Eastleigh Works, June 1995; locomotive included in sale when BREL works privatised; to Wimbledon Depot, for tyre turning, December 2000; returned to Eastleigh Works, January 2001; to LH Group, Barton under Needwood, for repairs, 14th October 2003; returned to Eastleigh Works, late 2003; to Alstom Transport,

Wolverton Works, 4th April 2006; to Chinnor & Princes Risborough Railway, Oxfordshire, for gala, 1st October 2013; to Knorr Bremse Rail Services Ltd, Wolverton Works, 22nd October 2013.

D3817 Horwich 1959 70D 8/89 P 08650

08650 to Foster Yeoman Quarries Ltd, Merehead Stone Terminal, Somerset, February 1989; to Foster Yeoman Quarries Ltd, Isle of Grain Stone Terminal, Kent, 7th May 1989; to Whatley Quarry, by 24th June 2001; to Foster Yeoman Quarries Ltd, Merehead Stone Terminal, 28th June 2002; to Foster Yeoman Quarries Ltd, Isle of Grain Stone Terminal, August 2002; to Wabtec, Doncaster, for repairs, about June 2003; to Foster Yeoman Quarries Ltd, Isle of Grain Stone Terminal, late 2003; to Foster Yeoman Quarries Ltd, Merehead Stone Terminal, for repairs, 25th February 2007; to Foster Yeoman Quarries Ltd, Isle of Grain Stone Terminal, 2nd March 2007; to Merehead Stone Terminal, 8th June 2012; to Knights Rail Services, Eastleigh Works, for repairs, 3rd July 2012; to Isle of Grain Stone Terminal, 2nd May 2013; to Arlington Fleet Services, Eastleigh Works, 19th March 2015; to Aggregate Industries, Isle of Grain Stone Terminal, 27th March 2015; to Moveright International, Wishaw, Warwickshire, for repairs, 11th December 2015; to Whatley Quarry, 2nd March 2016; to Merehead Stone Terminal, by May 2016; to Whatley Quarry, August 2016; to Arlington Fleet Services, Eastleigh Works, for overhaul, 20th March 2017; to Hanson, Whatley Quarry, 23rd August 2017; to Railway Support Services, Wishaw, for repairs, 27th March 2019.

D3819 Horwich 1959 86A 7/92 P 08652

08652 to Foster Yeoman Quarries Ltd, Merehead Stone Terminal, Somerset, 5th June 1993; to ARC (Southern) Ltd, Whatley Quarry, Somerset, March 1995; to Merehead Stone Terminal, 2001; to Whatley Quarry, 22nd October 2002; to Foster Yeoman Quarries Ltd, Acton Rail Terminal, London, 10th May 2004; to Whatley Quarry, October 2004; to Acton Rail Terminal, November 2004; to Whatley Quarry, February 2005; to Acton Rail Terminal, May 2005; to Whatley Quarry, by November 2005; to LH Group, Barton under Needwood, for overhaul, March 2008; to Whatley Quarry, about October 2008; to Traditional Traction, Wishaw, Warwickshire, for repairs, 14th May 2015; to Whatley Quarry, 2nd July 2015; to Merehead Stone Terminal, December 2015; to Whatley Quarry, February 2016; to Railway Support Services, Wishaw, on hire, 2nd December 2016; used at Axiom Rail, Wagon Works, Stoke on Trent, from 6th December 2016; returned to Whatley Quarry, 4th January 2017; to Merehead Stone Terminal, by 15th September 2017.

D3820 Horwich 1959 16A ? P 08653 / VERNON

08653 to Harry Needle Railroad Company, Long Marston; despatched from DB Schenker, Toton Depot, 18th January 2016.

D3822 Horwich 1959 40B 1/04 F 08655

08655 to T.J. Thomson Ltd, Stockton-on-Tees; despatched from EWS Thornaby Depot, 29th September 2005; to LH Group, Barton under Needwood, 16th June 2006; used for spares; remains scrapped by Donald Ward of Burton upon Trent, April 2007.

D3830 Horwich 1959 ? ? P 08663

08663 to Avon Valley Railway, Bitton; despatched from St Philip's Marsh Depot, Bristol, 8th July 2019; to Railway Support Services, Wishaw, for repairs, 23rd July 2019.

D3832 Crewe 1960 40B 55C F 08665

08665 to C.F. Booth Ltd, Rotherham, 16th December 2009; to Harry Needle Railroad Company (swapped for 08695); to Barrow Hill Engine Shed Society, Staveley, 17th December 2009; to European Metal Recycling, Kingsbury, for scrap, 4th October 2011; scrapped October 2011.

D3835 Crewe 1960 5A 9/95 F 08668

08668 to South Yorkshire Railway Preservation Society (HNRC), Meadowhall, Sheffield; despatched from Adtranz, Crewe, 21st March 1997; to Barrow Hill Engine Shed Society (HNRC), Staveley, 3rd November 1999; to Wabtec, Doncaster, 2nd July 2001; some of its parts used in the rebuild of 08517; to HNRC, Long Marston, 20th December 2007; used for spares, February 2011; remains to European Metal Recycling, Kingsbury, for scrap, 29th June 2011; scrapped on arrival.

D3836 Crewe 1960 9A 5/89 P 08669 / BOB MACHIN

08669 to Trafford Park Estates Ltd, Manchester, 28th March 1989; sold to Wabtec, Doncaster, November 2000; to First Great Western, Laira Depot, Plymouth, on hire, 27th April 2001; to Wabtec, Doncaster, 14th November 2002; to Freightliner, Port of Felixstowe, on hire, February 2003; to Wabtec, Doncaster, May 2003; to GNER, Bounds Green Depot, London, on hire, 28th May 2003; to Wabtec, Doncaster, about March 2005; to Tyseley Steam Depot, Birmingham, for tyre turning, 19th February 2014; returned to Wabtec, Doncaster; to LH Group, Barton under Needwood, 19th March 2015; to Wabtec, Doncaster, 1st May 2015; to Midland Road Depot, Leeds, for tyre turning, 18th May 2017; returned to Wabtec, Doncaster, 26th May 2017; to LH Group, Barton under Needwood, 31st May 2019.

D3837 Crewe 1960 66B 3/04 P 08670

08670 sold to T.J. Thomson Ltd, Stockton-on-Tees, January 2009; re-sold to Traditional Traction, Wishaw, Warwickshire, early 2009; re-sold to Colne Valley Enterprises Ltd, early 2009; to Colne Valley Railway, Castle Hedingham; despatched from DB Schenker, Motherwell Depot, 5th March 2009; to Pullman Rail, Canton Depot, Cardiff, on hire, 22nd April 2013; to Moveright International, Wishaw, Warwickshire, 11th July 2014; to Port of Felixstowe, on hire, 11th May 2015; to Moveright International, Wishaw, for repairs, 7th July 2015; to Chasewater Railway, Staffordshire, for testing, 11th December 2015; to Railway Support Services, Wishaw, 16th December 2015; to Bounds Green Depot, London, on hire, 4th February 2016; to Tyseley Depot, Birmingham, for tyre turning, 1st March 2019; to Bescot Yard, Walsall, on hire, 28th March 2019.

D3843 Horwich 1959 16A ? P 08676 / DAVE 2

08676 to Barrow Hill Engine Shed Society (HNRC), Staveley; despatched from DB Schenker, Toton Depot, 17th February 2016; to East Kent Railway, Shepherdswell, 12th October 2016.

D3845 Horwich 1959 41A 10/88 P 08678 / 555

08678 to Glaxochem Ltd, Ulverston, Cumbria, 4th May 1989; to Steamtown, Carnforth, Lancashire, for repairs, 3rd August 1990; returned to Glaxochem; to Steamtown, Carnforth, for repairs, 20th July 1992; returned to Glaxochem; sold to Steamtown, Carnforth, 8th November 1994; to Fragonset, Derby, for repairs, May 2001;

to Maintrain, Etches Park, on hire, by 9th June 2001; to Fragonset, Derby, by 3rd November 2001; to West Coast Railway Company, Carnforth, December 2001.

D3846 Horwich 1959 8J 6/76 F 08679

08679 to NCB, North Gawber Colliery, Mapplewell, Barnsley, 25th June 1976; to Royston Drift Mine, Barnsley, 7th September 1976; to North Gawber Colliery, Mapplewell, 5th April 1979; to C.F. Booth Ltd, Rotherham, for scrap, 18th April 1986; scrapped, April 1986.

D3849 Horwich 1959 ZF 6/95 P LIONHEART

08682 to ABB (Customer Support) Ltd, Doncaster Works, June 1995; locomotive included in sale when BRML works privatised; to Bombardier Transportation, Litchurch Lane Works, Derby, about January 2008; to Railway Vehicle Engineering Ltd, Railway Technical Centre, Derby, on hire, 31st July 2010; returned to Bombardier Transportation, Litchurch Lane Works, Derby, October 2010; to Railway Vehicle Engineering Ltd, Railway Technical Centre, Derby, for repairs, June 2013; to Bombardier Transportation, Litchurch Lane Works, Derby, 24th July 2013; to Railway Vehicle Engineering Ltd, Derby, for repairs, 22nd May 2014; to Bombardier Transportation, Litchurch Lane Works, Derby, 9th June 2014.

D3850 Horwich 1959 16A 2/04 P 08683

08683 to Traditional Traction, Wishaw, Warwickshire, 8th March 2007; to Gloucestershire Warwickshire Railway, Toddington, for storage, November 2009; to Freightliner, Port of Felixstowe, on hire, 10th February 2011; to Colne Valley Railway, Castle Hedingham, for storage, 17th August 2011; to Pullman Rail, Canton Depot, Cardiff, on hire, 16th April 2013; to EMD, Longport, Staffordshire, on hire, 3rd October 2013; to Traditional Traction, Wishaw, 2013; to Bombardier Transportation, Litchurch Lane Works, Derby, on hire, 23rd February 2015; to Traditional Traction, Wishaw, for overhaul, 1st April 2016; to Chasewater Railway, Staffordshire, for testing, 27th July 2016; to Epping Ongar Railway, Essex, for gala, 12th September 2016; to Bombardier Transportation, Litchurch Lane Works, Derby, 17th October 2016; to Railway Support Services, Wishaw, 27th January 2017; to Great Central Railway, Loughborough, 3rd May 2017; to Crown Point Depot, Norwich, on hire, 18th July 2017.

D3852 Horwich 1959 40B 2/09 P 08685

08685 to Barrow Hill Engine Shed Society (HNRC), Staveley; despatched from Immingham Depot, 19th September 2011; to Hope Cement Works, Derbyshire, on hire, 11th October 2016; to East Kent Railway, Shepherdswell, 5th December 2016.

D3854 Horwich 1959 ? 3/15 P 08995

08995 cut down cab locomotive; formerly numbered 08687 until 4th September 1987; to Railway Support Services, Wishaw, Warwickshire; despatched from Crewe Electric Depot, 7th September 2015.

D3859 Horwich 1959 5A 1/94 F 692

08692 to ABB Transportation, Crewe, 5th May 1994; to ABB (Customer Support) Ltd, Doncaster, about August 1995; returned to ABB Transportation, Crewe, by 17th August 1996; sold to Harry Needle Railroad Company, 2002; to West Coast Railway Company, Carnforth, July 2002; scrapped, May 2005.

D3861 Horwich 1959 81A 2/04 P 08694

08694 to C.F. Booth Ltd, Rotherham, 8th January 2009; to Great Central Railway, Loughborough, 22nd May 2009.

D3862 Horwich 1959 66B 2/04 F 08695

08695 to T.J. Thomson Ltd, Stockton-on-Tees; despatched from EWS Ayr Depot, 27th July 2007; to Barrow Hill Engine Shed Society (HNRC), Staveley, 11th September 2007; to C.F. Booth Ltd, Rotherham (swapped for 08665); scrapped December 2009.

D3863 Horwich 1959 ? ? P 08696

08696 acquired by Alstom Transport, Longsight Depot, Manchester, 2007; to Alstom Transport, Wembley Depot, London, 5th March 2012; to Arlington Fleet Services, Eastleigh Works, for repairs, 3rd August 2012; to Alstom Transport, Wembley Depot, London, 23rd August 2012; to Alstom Transport, Polmadie Depot, Glasgow, 14th December 2012; to Arlington Fleet Services, Eastleigh Works, for overhaul, 19th October 2015; to Alstom Transport, Wembley Depot, London, 26th February 2016.

D3864 Horwich 1959 16C 5/97 F 08697

08697 to Railway Vehicle Engineering Ltd, Railway Technical Centre, Derby; despatched from East Midlands Trains, Etches Park, Derby, July 2010; to D. Ward Ltd, Ilkeston, for scrap, 28th October 2014; scrapped, October 2014.

D3866 Horwich 1960 5A 10/93 F 08699 / 69

08699 to ABB Transportation, Crewe, 5th May 1994; sold to Cotswold Rail, Moreton in Marsh, early 2005; to Daventry International Rail Freight Terminal, on hire, about March 2005; to Cotswold Rail, May 2005; to Tyseley Steam Depot, Birmingham, for storage, 2005; to Allelys Ltd, Studley, Warwickshire, for storage, about 24th February 2006; sold to RMS Locotec, January 2007; to Corus, Shotton, for storage, 8th July 2009; to Weardale Railway, Wolsingham, 28th March 2013; scrapped on site by J. Denham Metals Ltd of Bishop Aukland, 23rd March 2018.

D3867 Horwich 1960 30A 9/93 P D3867

08700 to Barrow Hill Engine Shed Society (HNRC), Staveley, 27th June 2001; to West Coast Railway Company, Carnforth, 7th August 2002; to Bryn Engineering Ltd, Wigan, 22nd June 2004; to Embsay & Bolton Abbey Railway, 23rd June 2004; to East Lancashire Railway, Bury, about April 2007; to Barrow Hill Engine Shed Society, Staveley, 17th March 2014; to Bombardier Transportation, Ilford Depot, on hire, 10th June 2015.

D3868 Horwich 1960 16A ? P 08701 / TYNE 100

08701 to Harry Needle Railroad Company, Long Marston; despatched from DB Schenker, Toton Depot, 18th January 2016.

D3870 Horwich 1960 2F ? P 08703

08703 to Railway Support Services, Wishaw, Warwickshire; despatched from DB Cargo, Bescot Depot, 19th January 2017; to Network Rail, Springs Branch Depot, Wigan, on hire, 20th January 2017.

D3871 Horwich 1960 1E 11/89 P 08704

08704 despatched from BR Bletchley, 19th April 1990; to Port of Boston, 8th May

1990; to BR Doncaster Depot, for tyre turning, 20th January 1992; to Port of Boston, 29th January 1992; to Nene Valley Railway, Wansford, on loan, 3rd February 1993; to Port of Boston, 10th September 1997; to Wabtec, Doncaster, for repairs, 6th June 2001; to Port of Boston, 26th July 2001; sold to Harry Needle Railroad Company, June 2011; to East Lancashire Railway, Bury, 13th April 2012; sold to Riviera Trains Ltd of Crewe; to Crewe Diesel Depot, 18th September 2013; to Bombardier Transportation, Litchurch Lane Works, Derby, for repaint, 12th May 2014; to Riviera Trains Ltd, Crewe Diesel Depot, late 2014; to Nemesis Rail, Burton upon Trent, 23rd October 2015; to Ecclesbourne Valley Railway, Wirksworth, 5th March 2018.

D3873 Crewe 1960 5A ? P 08706
08706 sold to Harry Needle Railroad Company; despatched from Crewe Electric Depot; to Railway Support Services, Wishaw, Warwickshire, 31st August 2017.

D3874 Crewe 1960 55G 8/93 F 08707
08707 to South Yorkshire Railway Preservation Society (HNRC), Meadowhall, Sheffield; despatched from Adtranz, Crewe, 4th April 1997; to Barrow Hill Engine Shed Society, Staveley, 27th July 2001; to West Coast Railway Company, Carnforth, 11th July 2002; scrapped, February 2005.

D3876 Crewe 1960 2F ? P 08709
08709 to Railway Support Services, Wishaw, Warwickshire; despatched from Bescot Depot, September 2015; to Colne Valley Railway, Castle Hedingham, for storage, 3rd March 2016; to Railway Support Services, Wishaw, 26th May 2017.

D3878 Crewe 1960 ? ? P 08711
08711 to Harry Needle Railroad Company; despatched from Tees Yard, 2017; to Nemesis Rail, Burton upon Trent, 1st June 2017.

D3881 Crewe 1960 5A ? P 08714
08714 to Harry Needle Railroad Company; despatched from Crewe Electric Depot, October 2016; to Hope Cement Works, Derbyshire, on hire, 19th October 2016.

D3889 Crewe 1960 ? ? P 08721
08721 acquired by Alstom Transport, Longsight Depot, Manchester, 2007; to Alstom Transport, Edge Hill Depot, Liverpool, December 2011; to Arlington Fleet Services, Eastleigh Works, for repairs, 11th February 2015; to Alstom Transport, Wembley Depot, London, 30th November 2015; to Alstom Transport, Longsight Depot, Manchester, 11th January 2016; to East Lancashire Railway, Bury, for gala, 14th April 2016; to Alstom Transport, Longsight Depot, Manchester, by 22nd April 2016; to Alstom Transport, Technology Centre, Widnes, 1st August 2017.

D3892 Crewe 1960 ? ? P 08724
08724 to Great North Eastern Railway and allocated to Craigentinny Depot, Edinburgh; sold to RFS (Engineering) Ltd, Doncaster, 1997; to Foster Yeoman Quarries Ltd, Isle of Grain Stone Terminal, Kent, on hire, 9th September 1998; to ARC, Whatley Quarry, on hire, October 1998; to RFS (Engineering) Ltd, Doncaster, 11th January 1999; to Blue Circle, Hope Cement Works, Derbyshire, on hire, March 1999; to RFS (Engineering) Ltd, Doncaster, 1999; to ARC, Whatley Quarry, on hire, by 19th September

1999; to RFS (Engineering) Ltd, Doncaster, 10th June 2000; to Neville Hill Depot, Leeds, on hire, about September 2000; to Wabtec, Doncaster, for repairs, early 2003; returned to Neville Hill Depot, Leeds, on hire, by 19th June 2003; to Wabtec, Doncaster, for repairs, 10th January 2013; to Midland Road Depot, Leeds, for tyre turning, 2nd August 2017; returned to Wabtec, Doncaster, August 2017.

D3896 Crewe 1960 66B 11/87 F 08728

08728 to Deanside Transit Ltd, Glasgow, November 1987; to Harry Needle Railroad Company, Long Marston, 19th June 2007; to C.F. Booth Ltd, Rotherham, for scrap, 4th November 2009; scrapped, November 2009.

D3898 Crewe 1960 ZH 6/95 P 08730 / THE CALEY

08730 to Railcare Ltd, Springburn Works, Glasgow, June 1995; locomotive included in sale when BRML works privatised; to LH Group, Barton under Needwood, for repairs, 19th August 2009; returned to Springburn Works, Glasgow, about March 2010; to Midland Road Depot, Leeds, for tyre turning, 20th January 2017; to Knorr Bremse Rail Services Ltd, Springburn Works, Glasgow, 16th February 2017.

D3899 Crewe 1960 66B 12/95 F 08731

08731 to Swindon Works, for dual brake fitting, June 1983; found to have frame damage so identity changed with 08572; although really 08572, it emerged from Swindon Works numbered 08731; to T.J. Thomson Ltd, Stockton-on-Tees; despatched from EWS Motherwell Depot, 21st June 2002; to Foster Yeoman Quarries Ltd, Merehead Stone Terminal, 13th August 2003; to LH Group, Barton under Needwood, used for spares, about March 2004; remains returned to Merehead Stone Terminal, 16th April 2004; to Bodmin & Wenford Railway, Cornwall; used for spares, August 2008; to Merehead Stone Terminal, 29th August 2008; to J.W. Ransome & Sons, Frome, for scrap, March 2009; scrapped March 2009.

D3902 Crewe 1960 2F 8/96 F 08734

08734 to Dean Forest Railway, Lydney; despatched from Canton Depot, Cardiff, 17th June 2000; to Sims Metals Ltd, Newport, for scrap, 20th August 2011.

D3904 Crewe 1960 66B 11/87 F 08736 / 4

08736 to Deanside Transit Ltd, Glasgow, November 1987; to Harry Needle Railroad Company, Long Marston, 23rd May 2007; to C.F. Booth Ltd, Rotherham, for scrap, 5th November 2009; scrapped, November 2009.

D3905 Crewe 1960 5A ? P 08737

08737 to Locomotive Storage Ltd, Crewe Diesel Depot; despatched from Crewe Electric Depot, 21st January 2016.

D3906 Crewe 1960 16A ? P 08738

08738 to Colne Valley Railway, Castle Hedingham, for storage; despatched from DB Schenker, Toton Depot, 17th September 2015; to Railway Support Services, Wishaw, 24th April 2017; to GBRf, East Yard, Eastleigh, on hire, 30th March 2019.

D3908 Crewe 1960 ? 9/97 F 08740

08740 to T.J. Thomson Ltd, Stockton-on-Tees; despatched from Ferrybridge, about

October 2005; to LH Group, Barton under Needwood, 15th June 2006; used for spares; remains scrapped by Donald Ward of Burton upon Trent, April 2007.

D3910 Crewe 1960 81F ? P 08742

08742 to Harry Needle Railroad Company; despatched from Oxford; to Didcot Railway Centre, Oxfordshire, 25th May 2017; to East Kent Railway, Shepherdswell, 30th April 2018; to Barrow Hill Engine Shed Society, Staveley, 25th April 2019.

D3911 Crewe 1960 55H 3/93 P BRYAN TURNER

08743 to RFS (Engineering) Ltd, Doncaster, March 1993; to Grovehurst UK Paper Ltd, Sittingbourne, on hire, 25th January 1993; returned to RFS (Engineering) Ltd, Doncaster, by 1st November 1996; sold to ICI Billingham Works, Stockton-on-Tees, 27th January 1997; to ICI Wilton Works, Middlesbrough, April 2004; to Cleveland Potash Ltd, Teesport, on hire, about July 2005; returned to ICI Wilton Works, off hire, October 2005; to Hunslet Engine Company, Barton under Needwood, for repairs, 13th October 2011; to ICI Wilton Works, 2nd May 2012; to Wensleydale Railway, Leeming Bar, for gala, 13th September 2016; to Sembcorp Utilities Ltd, Wilton, 27th September 2016.

D3914 Crewe 1960 66B 2/99 F 08746

08746 to Barrow Hill Engine Shed Society, Staveley; despatched from Doncaster, 21st July 2003; used for spares to repair 08928; remains to C.F. Booth Ltd, Rotherham, for scrap, 31st August 2003; scrapped, January 2004.

D3918 Crewe 1960 81A 6/98 F 08750

08750 sold to RT Rail, Crewe, 2000; to Wabtec, Doncaster, for assessment; despatched from Stratford Depot, 7th August 2000; to Wessex Traincare Ltd, Eastleigh Works, on hire, 1st December 2000; returned to RT Rail, Crewe; to Ilford Depot, on hire, 29th January 2001; to Wabtec, Doncaster, for repairs, by 18th November 2001; to Channel Tunnel Rail Link, Beechbrook Farm, near Ashford, on hire, by March 2002; to Wabtec, Doncaster, 1st August 2002; to Wensleydale Railway, Leeming Bar, 21st July 2003; to Wabtec, Doncaster, December 2003; to Imreys Minerals, Quidhampton, Salisbury, on hire, 26th March 2004; to Wabtec, Doncaster, for repairs, about April 2005; returned to Quidhampton, about August 2005; to Wabtec, Doncaster, for repairs, 13th October 2005; to Tubelines, Ruislip, London, on hire, 16th May 2006; to First Capital Connect, Hornsey Depot, London, on hire, 30th May 2007; RT Rail acquired by RMS Locotec, 8th November 2007; to RMS Locotec, Wakefield, for repairs, about November 2007; returned to Hornsey Depot, London; to Bombardier Transportation, Ilford Depot, on hire, about April 2011; to Castle Cement Works, Ketton, on hire, November 2011; to RMS Locotec, Weardale Railway, Wolsingham, for repairs, 20th August 2012; to Castle Cement Works, Ketton, on hire, 20th November 2012; to RMS Locotec, Weardale Railway, Wolsingham, 1st February 2013; scrapped on site by Wanted Metal Recycling Ltd of Shildon, 18th December 2018.

D3919 Crewe 1960 41A 5/97 F 08751

08751 to RFS Engineering, Doncaster, by June 1998; used for spares; remains to C.F. Booth Ltd, Rotherham, January 2004; re-sold to RT Rail, Crewe; to Wabtec, Doncaster, for spares, April 2004; remains to C.F. Booth Ltd, Rotherham, for scrap, 7th July 2004; scrapped, 4th September 2004.

D3920 Crewe 1960 2F ? P 08752 / LENNY

08752 to Railway Support Services, Wishaw, Warwickshire; despatched from DB Cargo, Bescot Depot, January 2017; to Bombardier Transportation, Litchurch Lane Works, Derby, on hire, 13th January 2017; to Railway Support Services, Wishaw, for repairs, 22nd March 2017; to Gemini Rail Services, Wolverton Works, on hire, 5th February 2019.

D3922 Horwich 1961 60A 8/99 P 08754 / HO41

08754 to RT Rail, Crewe, August 1999; to Wabtec, Doncaster, autumn 1999; to Port of Felixstowe, on hire, about April 2000; to Freightliner, Garston Railport, Liverpool, on hire, 2000; to Freightliner, Dagenham Dock, on hire, by 12th November 2001; to Wabtec, Doncaster, for repairs, by 9th August 2002; to Grant Rail, March, on hire, 22nd April 2003; to Silverlink, Bletchley, on hire, 31st March 2004; to Reading PW Yard, by 24th November 2004; sold to RMS Locotec, Dewsbury, early May 2005; to PD Ports, Tees Dock, on hire, before 14th May 2005; to RMS Locotec, Dewsbury, 1st September 2005; to PD Ports, Tees Dock, on hire, 16th September 2005; to RMS Locotec, 25th November 2005; to Network Rail, Whitemoor Yard, March, on hire, 31st March 2006; to Bombardier Transportation, Ilford Depot, on hire, April 2008; to Network Rail, Whitemoor Yard, March, on hire, by July 2008; to Wabtec, Doncaster, October 2008; to Wabtec, Kilmarnock, April 2012; to Tyseley Steam Depot, Birmingham, for tyre turning, 21st September 2012; to Studley, for storage, September 2012; to North Norfolk Railway, Sheringham, 19th December 2012; to Crown Point Depot, Norwich, on hire, 20th December 2012; to Mid-Norfolk Railway, Dereham, 31st August 2017; to RMS Locotec, Weardale Railway, Wolsingham, 4th October 2018.

D3924 Horwich 1961 86A ? P 08756 / HO39

08756 to RT Rail, Crewe; despatched from Canton Depot, Cardiff, 9th September 2003; to Wabtec, Doncaster, 9th September 2003; to Bombardier Transportation, Doncaster, for storage, April 2004; to West Yard, Doncaster, for storage, July 2005; to Wabtec, Doncaster, for overhaul, 2006; to Brunner Mond, Northwich, on hire, September 2006; to RMS Locotec, on hire, 1st November 2006; used on Stirling to Alloa line contract; to Elsecar Steam Railway, near Barnsley, 14th November 2006; sold to RMS Locotec, Wakefield, about November 2007; to Network Rail, Whitemoor Yard, March, on hire, 2nd October 2008; to Corus, Shotton Steelworks, on hire, 6th April 2009; to Tyseley Steam Depot, Birmingham, for tyre turning, 28th January 2017; to Tata Steel, Shotton, 31st January 2017; to RMS Locotec, Weardale Railway, Wolsingham, 19th May 2017.

D3925 Horwich 1961 5A ? P 08757 / EAGLE

08757 to Telford Steam Railway, Shropshire; despatched from DB Cargo, Crewe Electric Depot, 16th January 2017; to Railway Support Services, Wishaw, on hire, June 2019 and moved direct to GBRf, South Terminal, Port of Felixstowe, on contract hire, 19th June 2019; to GBRf, Dagenham, on hire, 15th July 2019.

D3930 Horwich 1961 60A 8/99 P 08762 / OLD TOM

08762 to RT Rail, Crewe, August 1999; to Wabtec, Doncaster, September 1999; to Freightliner, Dagenham Dock, on hire, by 26th April 2000; to Port of Felixstowe, on hire, after 13th December 2001; to Channel Tunnel Rail Link, Beechbrook Farm, near Ashford, on hire, by February 2002; to Europort, Wakefield, on hire, May 2002; to Wabtec,

Doncaster, for repairs, by 10th June 2002; to Freightliner, Dagenham, on hire, about September 2002; to Imreys Minerals (ECC), Quidhampton, on hire, 16th January 2003; to Wabtec, Doncaster, for repairs, 30th March 2004; to Midland Road Depot, Leeds, on hire, by 28th July 2004; fitted with auto couplers, 2004; to Brunner Mond, Winnington Works, Northwich, on hire, about September 2004; suffered collision damage, early 2006; to Wabtec, Doncaster, for repairs, 19th April 2006; returned to Brunner Mond, Northwich, on hire; sold to RMS Locotec, about August 2007; to Cemex Rail Products, Washwood Heath, on hire, early August 2011; to Network Rail, Derby, on hire, 31st July 2015; to RMS Locotec, Weardale Railway, Wolsingham, 17th May 2017; to Locomotive Storage, Crewe, 28th February 2019.

D3932 Horwich 1961 64B 5/88 P 08764
08764 to RFS (Engineering) Ltd, Kilnhurst, May 1988; to ARC Ltd, Machen Quarry, near Newport, on hire, 26th September 1990; returned to RFS (Engineering) Ltd, 31st October 1990; to BREL Ltd, York, on hire, 1st November 1990; to Flixborough Wharf Ltd, Flixborough, Scunthorpe, on hire, 13th December 1990; to RFS (Engineering) Ltd, Doncaster, 8th February 1993; to Trans-Manche Link, Channel Tunnel (number 96), on hire, 10th February 1993; to RFS (Engineering) Ltd, Doncaster, June 1993; to Sheerness Steel Co Ltd, Sheerness, Kent, on hire, by 26th June 1993; returned to RFS (Engineering) Ltd, 1993; to Hartlepool Power Station, on hire, October 1993; returned to RFS (Engineering) Ltd, Doncaster, by 10th April 1994; to Flixborough Wharf Ltd, Flixborough, Scunthorpe, on hire, June 1994; returned to RFS (Engineering) Ltd, Doncaster, by 5th December 1996; to Transfesa UK Ltd, Riverside Terminal, Tilbury, on hire, 16th August 1997; to RFS (Engineering) Ltd, Doncaster, for repairs, by February 1998; to Transfesa Rail Terminal, Tilbury, by 13th January 1999; sold to Transfesa; resold to Alstom Transport, late 2012; to Alstom Transport, Stonebridge Park Heavy Repair Depot, London, 30th January 2013; to Alstom Transport, Polmadie Depot, Glasgow, 15th October 2015.

D3933 Horwich 1961 81A 5/08 P NPT
08765 to Harry Needle Railroad Company; despatched from DB Schenker, Eastleigh; to Boden Rail Engineering, Washwood Heath, 28th June 2011; to Nemesis Rail, Burton upon Trent, for storage, 24th November 2011; to Rail Restorations North East, Shildon, for repairs, 9th February 2015; to Barrow Hill Engine Shed Society, Staveley, 8th May 2015.

D3935 Horwich 1961 30A 1/94 P D3935
08767 to North Norfolk Railway, Sheringham; despatched from Colchester Depot, 22nd August 1994.

D3937 Derby 1960 87E 5/89 P D3937/ GLADYS
08769 to MoD Long Marston, Worcestershire, 20th April 1990; to The Fire Service College, Moreton in Marsh, 12th November 1991; to Dean Forest Railway, Lydney, 2nd March 2000; to Severn Valley Railway, Bridgnorth, 12th May 2003; to Dean Forest Railway, Lydney, March 2010.

D3940 Derby 1960 30A 1/94 P D3940
08772 to East Anglian Railway Museum, Wakes Colne, Essex; despatched from

Colchester Depot, by 27th March 1994; to North Norfolk Railway, Sheringham, 18th September 2001.

D3941 Derby 1960 16A 3/94 P D3941
08773 sold to Mike Darnall, Newton Heath, Manchester, 25th July 2000; to Embsay & Bolton Abbey Railway, February 2006.

D3942 Derby 1960 51L 10/88 P 08774
08774 to A.V. Dawson Ltd, Middlesbrough, September 1988; to Cobra Railfreight, Middlesbrough, on hire, October 1998; returned to A.V. Dawson Ltd, Middlesbrough, 1998; to Wabtec, Doncaster, for overhaul, late 2001; returned to A.V. Dawson Ltd, 22nd February 2002; to Moveright International, Wishaw, Warwickshire, for repairs, 26th October 2017; returned to A.V. Dawson Ltd, 13th July 2018.

D3948 Derby 1960 87E ? P 08780 / FRED
08780 to Cotswold Rail, Moreton in Marsh; despatched from Landore Depot, about June 2001; to East Lancashire Railway, Bury, November 2002; to Transplant Ltd (London Underground maintenance), West Ruislip, on hire, 26th April 2005; to Midland Road Depot, Leeds, 15th September 2005; to Wabtec, Doncaster, 27th March 2006; to L&NWR, Carriage Works, Crewe, 1st June 2006; sold to Locomotive Services Ltd; to Locomotive Services, Southall Depot, London, 21st December 2007; to Locomotive Services, Crewe, 11th June 2019.

D3950 Derby 1960 36A ? P 08782
08782 to Harry Needle Railroad Company; despatched from Doncaster Depot; to Barrow Hill Engine Shed Society, Staveley, June 2017.

D3951 Derby 1960 16A ? P 08783
08783 to European Metal Recycling, Kingsbury, Warwickshire; despatched from DB Schenker, Toton Depot, 4th August 2011.

D3952 Derby 1960 16A ? P 08784
08784 to Great Central Railway, Ruddington, Nottingham; despatched from DB Cargo, Toton Depot, 14th December 2016.

D3953 Derby 1960 86A 3/89 P 004 / CLARENCE
08785 to RFS (Engineering) Ltd, Kilnhurst, 25th September 1990; to Grovehurst Paper Ltd, Sittingbourne, Kent, on hire, 4th June 1991; to Channel Tunnel Rail Link, on hire, 11th September 1992; to RFS (Engineering) Ltd, Doncaster, July 1993; to BASF Chemicals Ltd, Seal Sands, on hire, 9th August 1994; returned to RFS (Engineering) Ltd, Doncaster, 25th November 1994; sold to Freightliner, 14th July 1997.

D3954 Derby 1960 16A 12/08 P 08786
08786 to Harry Needle Railroad Company; to Barrow Hill Engine Shed Society, Staveley; despatched from DB Schenker, Doncaster Depot, 24th January 2011.

D3955 Derby 1960 86A 2/91 P 08296
08787 to ABB Transportation, Crewe, February 1991; overhauled, 1992; given identity 001, by October 1992: to ABB British Wheelset Ltd, Trafford Park, Manchester, January

1995; to Adtranz, Crewe, August 1996; to East Somerset Railway, Cranmore, for storage, February 2000; to Foster Yeoman Quarries Ltd, Merehead Stone Terminal, about March 2000; to Hanson Aggregates, Whatley Quarry, by 1st October 2000; number 08296 applied, late 2000; to Foster Yeoman Quarries Ltd, Isle of Grain Stone Terminal, 28th June 2002; to Foster Yeoman Quarries Ltd, Whatley Quarry, 29th July 2002; to Merehead Stone Terminal, by 28th June 2003; to Whatley Quarry, by 17th April 2004; to Machen Quarry, near Newport, by 27th May 2004; to Whatley Quarry, by 12th June 2004; to Acton Rail Terminal, late 2004; to Isle of Grain Stone Terminal, by February 2005; to Merehead Stone Terminal, by 11th July 2005; to Hanson Quarry Products, Machen Quarry, near Newport, 6th April 2006; to Whatley Quarry, 27th March 2019.

D3956 Derby 1960 16C 1/94 P 08788

08788 to Great Central Railway, Loughborough, 30th March 1994; to RT Rail, Crewe, 24th February 1999; to Scot Rail, Inverness, on hire, 1st September 1999; to Manchester Ship Canal Company, Barton Dock, on hire, 21st April 2005; to Wabtec, Doncaster, for repairs, 2nd June 2005; to Scot Rail, Inverness, on hire, 16th November 2005; sold to RMS Locotec, Wakefield, about November 2007; to Craigentinny Depot, Edinburgh, for tyre turning, 19th December 2012; to Scot Rail, Inverness, 20th December 2012; to RMS Locotec, Weardale Railway, Wolsingham, 16th November 2015; to Tata Steel, Shotton, on hire, 17th August 2016; to PD Ports, Tees Dock, on hire, 10th July 2019.

D3958 Derby 1960 ? ? P 08790

08790 acquired by Alstom Transport, Longsight Depot, Manchester, 2007; to Alstom Transport, Oxley Depot, Wolverhampton, 26th April 2007; to Arlington Fleet Services, Eastleigh Works, for overhaul, 18th June 2014; to Alstom Transport, Oxley Depot, Wolverhampton, 11th February 2015; to Arlington Fleet Services, Eastleigh Works, for repairs, 14th December 2016; to Alstom Transport, Edge Hill Depot, Liverpool, about October 2017.

D3963 Derby 1960 87E ? P 08795

08795 to Llanelli & Mynydd Mawr Railway, Cynheidre; despatched from Landore Depot, 22nd March 2019.

D3966 Derby 1960 16A 11/09 P 08798

08798 to European Metal Recycling, Kingsbury, 15th August 2011; to European Metal Recycling, Attercliffe, Sheffield, early July 2012.

D3967 Derby 1960 82D ? P 08799 / FRED

08799 to Harry Needle Railroad Company; despatched from Westbury; to East Kent Railway, Shepherdswell, 8th June 2017.

D3969 Derby 1960 86A 6/00 F 08801

08801 sold to RT Rail, Crewe; to Wabtec, Doncaster, 8th September 2003; to C.F. Booth Ltd, Rotherham, for scrap, February 2004; scrapped, 24th March 2004.

D3970 Derby 1960 16A ? P 08802

08802 sold to Harry Needle Railroad Company; to Railway Support Services, Wishaw, Warwickshire; despatched from DB Schenker, Toton Depot, 18th February 2016.

D3972 Derby 1960 5A ? P 08804

08804 to Harry Needle Railroad Company; despatched from Crewe Electric Depot; to Moveright International, Wishaw, Warwickshire, 1st September 2017; to East Kent Railway, Shepherdswell, 4th September 2017.

D3975 Derby 1960 66B 6/04 F 08807

08807 to T.J. Thomson Ltd, Stockton-on-Tees, 26th April 2007; to EWS Thornaby Depot, for tyre turning, 30th October 2007; to A.V. Dawson Ltd, Middlesbrough, for spares, week-ending 23rd November 2007; remains to C.J. Prosser Ltd, Cargo Fleet, for scrap, 7th January 2017; scrapped January 2017.

D3977 Derby 1960 8J 12/93 P 08809 / 24

08809 to Otis Euro Trans Rail Ltd, Salford, Manchester, December 1993; to Flixborough Wharf Ltd, Flixborough, Scunthorpe, on hire, week-ending 12th January 1996; returned to Otis Euro Trans Rail Ltd; sold to Harry Needle Railroad Company, late 1999; to Fragonset, Derby, for certification, about March 2000; to Barrow Hill Engine Shed Society (HNRC), Staveley, 13th June 2000; to Freightliner, Coatbridge, on hire, 29th June 2000; to Motherwell Depot, for repairs, September 2001; returned to Freightliner, Coatbridge, on hire; to Barrow Hill Engine Shed Society, Staveley, 24th May 2002; sold to Cotswold Rail, Moreton in Marsh, 29th June 2002; to Anglia Railways, Crown Point Depot, Norwich, on hire, 24th October 2002; to Ilford Depot, for tyre turning, 10th December 2003; returned to Crown Point Depot, Norwich, 17th December 2003; to Brush, Loughborough, for overhaul, 17th March 2005; to Allelys Ltd, Studley, Warwickshire, for storage, 7th February 2006; sold to RMS Locotec, Wakefield, October 2007; to Corus, Shotton, for storage, 26th October 2009; to Boden Rail Engineering, Washwood Heath, 29th November 2010; to RMS Locotec, Weardale Railway, Wolsingham, 12th November 2013; to PD Ports, Tees Dock, on hire, 27th August 2014; to Tata Steel, Shotton, on hire, June 2017; to Hanson Cement, Ketton Cement Works, on hire, week-ending 11th May 2019.

D3978 Derby 1960 NC ? P 08810

08810 to Cotswold Rail, Moreton in Marsh, about June 2001; to Brush, Loughborough, for overhaul, 15th August 2001; to Anglia Railways, Crown Point Depot, Norwich, on hire, 18th September 2001; to Railway Age, Crewe, 10th December 2003; to London & North Western Railway Company, Carriage Works, Crewe, about March 2004; to Arlington Fleet Services, Eastleigh Works, December 2011; to Heaton Depot, Newcastle upon Tyne, 25th May 2012; to L&NWRC, Traction and Rolling Stock Depot, Eastleigh, 9th December 2014.

D3981 Derby 1960 51L 1/00 F 08813

08813 to Harry Needle Railroad Company, Long Marston; despatched from EWS Thornaby Depot, 26th September 2006; used for spares, January 2011; remains to T.J. Thomson Ltd, Stockton-on-Tees, for scrap, 4th February 2011; scrapped, 9th February 2011.

D3984 Derby 1960 51L 2/86 F HO25

08816 to Cobra Railfreight Ltd, Middlesbrough, 15th February 1986; to Harry Needle Railroad Company, 1995; to Johnson Ltd, Widdrington Disposal Point, Northumberland, on hire, 19th August 1995; to RFS (Engineering) Ltd, Doncaster, 10th July 1998; scrapped, August 1999.

D3986 Derby 1960 5A 4/97 P 08818 / 4 / MOLLY
08818 to Railway Age, Crewe, September 1997; sold to Harry Needle Railroad Company, March 1997; to London & North Western Railway Company, Carriage Works, Crewe, on hire, by June 1998; to Port of Felixstowe, on hire, 13th August 1999; to Barrow Hill Engine Shed Society (HNRC), Staveley, 3rd November 1999; to Freightliner, Basford Hall, Crewe, on hire, 10th June 2000; to Barrow Hill Engine Shed Society, Staveley, 12th April 2001; to Freightliner, Stourton, Leeds, on hire, 24th January 2002; to Battlefield Line, Shackerstone, 21st August 2002; to Freightliner, Coatbridge, on hire, 4th November 2002; to Battlefield Line, Shackerstone, 4th December 2002; to Freightliner, Calvert Landfill Site, on hire, 12th May 2003; to MoD Bicester, June 2004; to Severn Valley Railway, Bridgnorth, on hire, 21st September 2004; to Network Rail, Whitemoor Yard, March, on hire, about October 2004; to HNRC, Long Marston,12th October 2005; to Barrow Hill Engine Shed Society, Staveley, 21st March 2006; to Flixborough Wharf Ltd, Flixborough, Scunthorpe, on hire, 20th May 2007; to GBRf, Trafford Park, Manchester, on hire, 30th May 2014; to Celsa, Cardiff, on hire, 16th July 2015; to Barrow Hill Engine Shed Society, Staveley, 10th January 2017; to Midland Road Depot, Leeds, for tyre turning, 27th January 2017; to Barrow Hill Engine Shed Society, Staveley, 3rd February 2017; to Celsa, Cardiff, on hire, 26th June 2017; to Barrow Hill Engine Shed Society, Staveley, July 2017; to GBRf, c/o Ford Motor Co Ltd, Speke, Liverpool, on hire, 19th July 2017.

D3987 Derby 1960 86A 9/99 F 08819
08819 to RT Rail, Crewe, September 2003; to Wabtec, Doncaster, for storage, about November 2003; to Bombardier Transportation, Doncaster, for storage, April 2004; to West Yard, Doncaster, for storage, by July 2005; to C.F. Booth Ltd, Rotherham, for scrap, 21st May 2008; scrapped, August 2008.

D3991 Derby 1960 ZF 6/95 P 08823 / KEVLA
08823 to ABB (Customer Support) Ltd, Doncaster Works, June 1995; locomotive included in sale when BRML works privatised; to Churnet Valley Railway, Cheddleton, Staffordshire, November 2000; to LH Group, Barton under Needwood, for repairs, 3rd August 2007; purchased by Hunslet Engine Company, March 2008; to Manchester Ship Canal Company, Trafford Park, Manchester, on hire, 24th March 2008; to LH Group, Barton under Needwood, for repairs, 22nd May 2009; to Sheerness Steel Co Ltd, Sheerness, Kent, on hire, 8th June 2010; to LH Group, Barton under Needwood, for repairs, 13th August 2010; to Sheerness Steel Co Ltd, Sheerness, Kent, on hire, 20th August 2010; to LH Group, Barton under Needwood, 15th February 2012; to Daventry International Rail Freight Terminal, on hire, 27th February 2012; to LH Group, Barton under Needwood, for repairs, 20th February 2015; returned to Daventry International Rail Freight Terminal, on hire, 6th March 2015; to LH Group, Barton under Needwood, 19th October 2017; to Midland Road Depot, Leeds, for tyre turning, 13th December 2018; to LH Group, Barton under Needwood, 17th December 2018; to Tata Steel, Shotton, on hire, 18th January 2019.

D3992 Derby 1960 5A ? P IEMD 01

08824 to Barrow Hill Engine Shed Society (HNRC), Staveley; despatched from DB Schenker, Crewe Depot, 4th December 2015.

D3993 Derby 1960 81A 12/99 P 08825

08825 to Battlefield Line, Shackerstone; despatched from EWS Springs Branch Depot, Wigan, 7th October 2005; to Chinnor & Princes Risborough Railway, Oxfordshire, 14th May 2013.

D3994 Derby 1960 66B 12/95 F 08826

08826 to T.J. Thomson Ltd, Stockton-on-Tees; despatched from EWS Motherwell Depot, July 2002; to East Somerset Railway, Cranmore, 8th October 2003; to Whatley Quarry, about May 2004; to Foster Yeoman Quarries Ltd, Merehead Stone Terminal, 9th June 2004; to Whatley Quarry, February 2005; to Merehead Stone Terminal, early June 2005; to Mid Hants Railway, Ropley, for storage, March 2009; to Knight's Rail Services, Eastleigh Works, 21st October 2010; used for spares; remains scrapped, April 2011.

D3995 Derby 1960 66B 3/00 F 08827

08827 to Barrow Hill Engine Shed Society (HNRC), Staveley; despatched from EWS Motherwell Depot, 2nd September 2005; to HNRC, Long Marston, week-ending 28th July 2006; used for spares, January 2011; remains to European Metal Recycling, Kingsbury, for scrap, 7th July 2011; scrapped September 2011.

D3997 Derby 1960 16A 6/93 F 08829

08829 to European Metal Recycling, Kingsbury, August 2000; to Barrow Hill Engine Shed Society, Staveley, 19th March 2001; to West Coast Railway Company, Carnforth, 7th October 2002; scrapped, February 2005.

D3998 Derby 1960 ? ? P 08830

08830 to East Somerset Railway, Cranmore (on hire from Cardiff Railways Ltd and despatched from Cardiff Cathays), 2nd October 1996; to Foster Yeoman Quarries Ltd, Torr Works, on hire, July 1997; returned to East Somerset Railway, about August 1997; may have worked at Merehead Stone Terminal for periods in this era before returning to Cardiff Railways Ltd, off lease, 18th September 1999; to L&NWR Ltd, Carriage Works, Crewe, on hire, 30th December 1999; to Crewe Works, for open day, 19th May 2000; returned to L&NWR Ltd, Crewe, May 2000; to Freightliner, Port of Felixstowe, on hire, about 12th April 2005; to Heritage Centre, Crewe, 13th July 2005; to Midland Road Depot, Leeds, on hire, by 16th August 2005; to Wabtec, Doncaster, 2007; to Heritage Centre, Crewe, 2007; to Peak Rail, Rowsley, 5th November 2015.

D4002 Derby 1960 ? ? P 08834

08834 to Transmanche-Link, Dolland Moor, Kent, on hire, 23rd September 1992; to BR Stratford Depot, London, 23rd November 1992; sold to RFS (Engineering) Ltd, Doncaster, 27th April 1993; to GNER, Bounds Green Depot, London, on hire, 1997; returned to RFS (Engineering) Ltd, Doncaster, by 14th February 1998; to GNER, Bounds Green Depot, London, on hire, 2nd August 1999; to Wabtec, Doncaster, 28th May 2003; to Foster Yeoman Quarries Ltd, Merehead Stone Terminal, on hire, 19th January 2004; to Daventry International Rail Freight Terminal, on hire, November 2004; to Wabtec,

Doncaster, 26th May 2005; to Foster Yeoman Quarries Ltd, Merehead Stone Terminal, on hire, early 2006; returned to Wabtec, Doncaster, about April 2006; sold to Direct Rail Services, May 2006; to DRS, Kingmoor Depot, Carlisle, 2nd October 2006; to DRS, Gresty Bridge Depot, Crewe, 23rd January 2007; sold to Harry Needle Railroad Company, January 2009; to Basford Hall, Crewe, 22nd May 2009; to Serco, Old Dalby Test Centre, Leicestershire, 27th May 2009; to HNRC, Barrow Hill Engine Shed Society, Staveley, for overhaul, 10th March 2015; to GBRf, Trafford Park, Manchester, on hire, 5th January 2016; to Barrow Hill Engine Shed Society, Staveley, March 2016; to GBRf, Dagenham, on hire, 6th May 2016; to GBRf, Allerton Depot, Liverpool, on hire, 11th August 2016.

D4014 Horwich 1961 8J 9/89 P 003
08846 to ABB Transportation, Litchurch Lane Works, Derby, October 1989; to ABB Transportation, Crewe, October 1993; to ABB Transportation, York, September 1995; to ABB Transportation, Crewe, for overhaul, 30th August 1996; to Adtranz, Litchurch Lane Works, Derby, 15th May 1998; to Fragonset, Derby, for overhaul, about March 2004; to Bombardier Transportation, Litchurch Lane Works, Derby, about June 2004; to Railway Support Services, Wishaw, Warwickshire, for repairs, 9th April 2015; repairs delayed; sold to Railway Support Services, Wishaw, Warwickshire, 2018; to St Philip's Marsh Depot, Bristol, for tyre turning, 21st June 2019; to Railway Support Services, Wishaw, 24th June 2019; to East Midlands Trains, Neville Hill Depot, Leeds, on hire, 15th July 2019.

D4015 Horwich 1961 ZG 6/95 P 08847
08847 to Wessex Traincare Ltd, Eastleigh Works, June 1995; locomotive included in sale when BREL works privatised; sold to Cotswold Rail, Moreton in Marsh, 18th May 2001; to Brush, Loughborough, for overhaul, 24th July 2001; to Cotswold Rail, about July 2002; to Anglia Railways, Crown Point Depot, Norwich, on hire, September 2002; to GBRf, Railport, Doncaster, on hire, November 2002; to Wabtec, Doncaster, 28th March 2003; to British Gypsum, Mountfield, East Sussex, on hire, 7th May 2003; to Anglia Railways, Crown Point Depot, Norwich, on hire, 11th September 2003; to Horton Road Depot, Gloucester, for storage, 2006; purchased by RMS Locotec, Wakefield, October 2007; to Wabtec, Doncaster, early 2008; to Anglia Railways, Crown Point Depot, Norwich, on hire, by 26th March 2008; to Allelys, Studley, Warwickshire, 20th December 2012; to Boden Rail Engineering, Washwood Heath, week-ending 28th February 2014; to Tyseley Steam Depot, Birmingham, for tyre turning, 14th April 2014; to Boden Rail Engineering, Washwood Heath, 18th April 2014; to East Coast Trains, Bounds Green Depot, London, on hire, 3rd December 2014; to Mid-Norfolk Railway, Dereham, 2nd July 2015; to Crown Point Depot, Norwich, on hire, 4th July 2015; to Mid-Norfolk Railway, Dereham, 31st August 2017; to PD Ports, Tees Dock, on hire, 14th May 2019.

D4018 Horwich 1961 81D 12/92 P 08850
08850 to West Somerset Railway, Minehead, 13th September 1993; to North Yorkshire Moors Railway, Grosmont, 11th March 1998.

D4021 Horwich 1961 EC 1/96 P 08853
08853 to RFS (Engineering) Ltd, Doncaster, 30th January 1997; overhauled; to Great North Eastern Railway, Bounds Green Depot, London, on hire, about March 1997; returned to RFS (Engineering) Ltd, Doncaster, 1998; to GNER, Bounds Green Depot,

London, on hire, June 2003; to Wabtec, Doncaster, 8th February 2008; to Midland Road Depot, Leeds, for tyre turning, 7th December 2018; to Wabtec, Doncaster, 12th December 2018.

D4033 Darlington 1960 5A ? P 08865 / GILLEY
08865 purchased by Harry Needle Railroad Company; to Moveright International, Wishaw, Warwickshire, for storage; despatched from Crewe Electric Depot, 30th November 2015; to Barrow Hill Engine Shed Society, Staveley, 20th October 2016; to Hope Cement Works, Derbyshire, on hire, 16th February 2017.

D4035 Darlington 1960 8F ? F HL1007
08867 to RMS Locotec, Dewsbury, about 1993; to Brunner Mond, Northwich, on hire, 6th January 1994; to RMS Locotec, by March 1998; to Brunner Mond, on hire, 1998; to RMS Locotec, Dewsbury, for repairs, 29th November 1999; to Marcroft Wagon Works, Horbury, Wakefield, on hire, early 2000; to RMS Locotec, Dewsbury, by February 2001; to Cobra Railfreight, Wakefield, on hire, 2nd July 2002; to EWS, Ferrybridge Depot, for storage, September 2002; to T.J. Thomson Ltd, Stockton-on-Tees, for scrap, 4th October 2005; scrapped, 20th June 2007.

D4036 Darlington 1960 31B 12/92 P 08868
08868 to South Yorkshire Railway Preservation Society (HNRC), Meadowhall, Sheffield, 22nd February 1994; to East Lancashire Railway, Bury, 16th April 1994; to RFS (Engineering) Ltd, Doncaster, on hire, 3rd September 1997; sub-hired to Fastline Track Renewals, Peterborough, 12th September 1997; used at May-Gurney, Connington Tip, near Peterborough; to Railway Age, Crewe, 2nd June 1998; to Freightliner, Basford Hall, Crewe, on hire, January 1999; to MoD, Long Marston, on hire, 2nd February 1999; to Railway Age, Crewe, 6th April 1999; to Port of Felixstowe, on hire, 6th August 1999; to L&NWR, Carriage Works, Crewe, on hire, 13th August 1999; to Freightliner, Basford Hall, Crewe, on hire, 18th September 2000; to Freightliner, Trafford Park, Manchester, on hire, 30th April 2001; to Barrow Hill Engine Shed Society, Staveley, 17th April 2002; to Blue Circle, Hope Cement Works, Derbyshire, on hire, 15th November 2002; to Port of Felixstowe, on hire, 25th July 2003; to Midland Road Depot, Leeds, on hire, 7th December 2003; to Port of Felixstowe, on hire, 28th July 2004; to London & North Western Railway Company Ltd, Carriage Works, Crewe, 24th November 2004.

D4037 Darlington 1960 NC ? F 08869
08869 sold to Cotswold Rail, Moreton in Marsh, about June 2001; to Mid-Norfolk Railway, Dereham, for loading; despatched from Crown Point Depot, Norwich, 14th August 2001; to Brush, Loughborough, for repairs, 15th August 2001; to Barrow Hill Engine Shed Society (HNRC), Staveley, 27th August 2003; to HNRC, Long Marston, for storage, September 2006; to European Metal Recycling, Kingsbury, for scrap, 29th September 2010; scrapped, January 2011.

D4038 Darlington 1960 55G 5/93 P HO24
08870 to South Yorkshire Railway Preservation Society, Meadowhall, Sheffield, 1st March 1994; to Cobra Railfreight, Wakefield, on hire, 9th December 1994; to SYRPS, Sheffield, 15th October 1997; sold to RMS Locotec, Dewsbury, September 1998; to Anglian Railways, Crown Point Depot, Norwich, on hire, 26th September 1998; returned

to RMS Locotec, for repairs, 30th March 1999; to Anglian Railways, Crown Point Depot, Norwich, on hire, February 2000; to RMS Locotec, for repairs, 1st March 2001; to Anglian Railways, Crown Point Depot, Norwich, on hire, about March 2001; to RMS Locotec, 14th June 2001; to Bombardier Transportation, Horbury, Wakefield, on hire, 19th June 2001; to Redland, Barrow upon Soar, on hire, by 12th January 2002; to Ford, Bridgend, on hire, March 2002; to Redland, Barrow upon Soar, on hire, 6th November 2002; to RMS Locotec, 13th October 2004; to Castle Cement Works, Ketton, on hire, 27th January 2005; to Weardale Railway, Wolsingham, on hire, (used at Colas coal loading siding), 6th July 2010; to Wabtec, Kilmarnock, on hire, 16th February 2015; to RMS Locotec, Weardale Railway, Wolsingham, 9th July 2015; to Tata Steel, Trostre Works, Llanelli, on hire, 7th December 2015; to Calkeld Heavy Haulage, Stourton, Leeds, for storage, 26th April 2016; to Castle Cement Works, Ketton, on hire, 7th November 2016; to RMS Locotec, Weardale Railway, Wolsingham, for repairs, 5th July 2018; sold to Eastern Railway Services, about June 2019.

D4039 Darlington 1960 41A 10/90 P 08871 / HO74

08871 to Humberside Sea & Land Services Ltd, Royal Dock, Grimsby; despatched from BR Immingham Depot, 16th December 1990; sold to Cotswold Rail, Moreton in Marsh, by 7th April 2001; to Brush, Loughborough, for repairs, 18th April 2001; to Anglia Railways, Crown Point Depot, Norwich, on hire, August 2001; to Wabtec, Doncaster, for repairs, 5th November 2002; to Bombardier Transportation, Ilford, for tyre turning, January 2004; to Network Rail, Whitemoor Yard, March, on hire, 1st April 2004; to Daventry International Rail Freight Terminal, on hire, 8th October 2004; to Brush, Loughborough, for repairs, 9th November 2004; returned to Daventry International Rail Freight Terminal, 14th December 2004; to Brush, Loughborough, for repairs, 11th April 2005; to Anglia Railways, Crown Point Depot, Norwich, on hire, 28th November 2005; to Horton Road Depot, Gloucester, for storage, about August 2006; purchased by RMS Locotec, Wakefield, October 2007; to Anglia Railways, Crown Point Depot, Norwich, on hire, by 26th May 2007; to Wabtec, Doncaster, for repairs, by August 2008; to East Coast Trains, Craigentinny Depot, Edinburgh, on hire, 27th November 2010; to Wabtec, Doncaster, 13th September 2011; to Boden Rail Engineering, Washwood Heath, for repairs, December 2011; to Cemex Rail Products, Washwood Heath, on hire, about February 2012; to PD Ports, Tees Dock, on hire, 16th April 2012; to RMS Locotec, Weardale Railway, Wolsingham, for repairs, 22nd August 2012; to PD Ports, Tees Dock, on hire, 4th December 2012; to RMS Locotec, Weardale Railway, Wolsingham, for repairs, 1st August 2014; to Tata Steel, Trostre Works, Llanelli, on hire, 26th April 2016; to Bombardier Transportation, Ilford Depot, on hire, 26th February 2018.

D4040 Darlington 1960 40B 2/04 P 08872

08872 to European Metal Recycling, Attercliffe, Sheffield; despatched from DB Schenker, Immingham Depot, 20th August 2010.

D4041 Darlington 1960 5A ? P 08873

08873 to ABB Transportation, Derby, 15th May 1998; sold to RT Rail, Crewe, about April 2000; to RFS (Engineering) Ltd, Doncaster, for repairs, May 2000; to L&NWR Ltd, Carriage Works, Crewe, on hire, by 7th September 2000; to Manchester Ship Canal Company, Barton Dock, on hire, May 2005; purchased by Hunslet Engine Company, Barton under Needwood, November 2005 (remained on hire to MSC Barton Dock); to

LH Group, Barton under Needwood, for repairs, 20th June 2006; to Manchester Ship Canal Company, Barton Dock, on hire, 14th November 2006; to LH Group, Barton under Needwood, for repairs, 24th April 2007; to Port of Felixstowe, on hire, 22nd October 2007; to Manchester Ship Canal Company, Barton Dock, Manchester, on hire, 7th January 2009; to Freightliner, Trafford Park, Manchester, on hire, May 2009; to LH Group, Barton under Needwood, for repairs, 20th August 2009; to Innovative Logistics, Brierley Hill, on hire, November 2009; to Freightliner, Southampton Docks, on hire, 20th May 2010; to LH Group, Barton under Needwood, for repairs, 15th February 2011; to Freightliner, Southampton Docks, on hire, week commencing 4th July 2011; to LH Group, Barton under Needwood, for repairs, 23rd November 2012; to Freightliner, Southampton Docks, on hire, 22nd January 2013; to Hams Hall Rail Freight Terminal, Coleshill, on hire, 20th May 2013; to LH Group, Barton under Needwood, 23rd August 2016; to Freightliner, Southampton Docks, on hire, 19th October 2016; to LH Group, Barton under Needwood, 8th May 2017.

D4042 Darlington 1960 55H 2/92 P 08874

08874 to RFS (Engineering) Ltd, Kilnhurst; 2nd July 1992; to Trans-Manche Link, Channel Tunnel contract (number 97), on hire, 13th February 1993; returned to RFS (Engineering) Ltd, Doncaster, 11th February 1994; to Teesbulk Handling, Middlesbrough, on hire, 2nd June 1994; returned to RFS (Engineering) Ltd, Doncaster, 13th September 1994; to Sheerness Steel Co Ltd, Sheerness, Kent, on hire, 16th September 1995; returned to RFS (Engineering) Ltd, Doncaster, May 1997; purchased by RT Rail, Crewe, October 1998; to Crewe Depot, for certification, early October 1998; to Hays Chemicals, Sandbach, on hire, 12th October 1998; to RFS (Engineering) Ltd, Doncaster, for repairs, 8th January 1999; to Silverlink, Bletchley, on hire, 26th February 1999; to LH Group, Barton under Needwood, for fitting with Train Protection Warning System equipment, early April 2004; to Silverlink, Bletchley, on hire, August 2004; sold to RMS Locotec, Wakefield, about November 2007; to Dartmoor Rail, Meldon Quarry, on hire, 19th December 2007; to Mid-Norfolk Railway, Dereham, 14th April 2008; to Anglia Railways, Crown Point Depot, Norwich, on hire, late April 2008; to North Norfolk Railway, Sheringham, 20th December 2012; to Allelys, Studley, Warwickshire, 21st December 2012; to Anglia Railways, Crown Point Depot, Norwich, on hire, about October 2013; to Tata Steel, Shotton Works, on hire, 9th July 2015; to RMS Locotec, Weardale Railway, Wolsingham, 7th August 2018.

D4043 Darlington 1960 51L 5/91 F 08875

08875 to RFS (Engineering) Ltd, Kilnhurst, 20th November 1991; used for spares; remains scrapped on site by C.F. Booth Ltd, Rotherham, August 1993.

D4044 Darlington 1960 36A 10/91 F 08876

08876 to RFS (Engineering) Ltd, Kilnhurst; despatched from BR Tinsley Depot, 2nd July 1992; to RFS (Engineering) Ltd, Doncaster, autumn 1993; used for spares; remains scrapped on site by Hudson Ltd of Madeley, Telford, April 1994.

D4045 Darlington 1961 8F ? P WIGAN 1 / REVENGE

08877 purchased by Harry Needle Railroad Company; to Barrow Hill Engine Shed Society, Staveley; despatched from Springs Branch Depot, Wigan, 29th October 2015; to Celsa, Cardiff, on hire, 6th June 2019.

D4047 Darlington 1961 87B ? P 08879

08879 to Raxstar Ltd, Eastleigh Works; despatched from DB Cargo, Margam Depot, 12th December 2016; sold to Harry Needle Railroad Company, March 2018; to Traditional Traction, Wishaw, Warwickshire, for repairs, 4th May 2018; to Barrow Hill Engine Shed Society, Staveley, 29th June 2018.

D4056 Darlington 1961 40B 6/72 F D4056 / 55

to NCB Ashington Central Workshops, January 1973; to Shilbottle Colliery, February 1973; to Ashington Central Workshops, March 1974; to Shilbottle Colliery, 18th June 1974; scrapped on site by T.J. Thomson Ltd of Stockton-on-Tees, March 1983.

D4067 Darlington 1961 41J 12/70 P 10119

to NCB Betteshanger Colliery, Kent, April 1971; seen at Betteshanger numbered 1802/B4, 15th October 1972; to Snowdown Colliery, Kent, 27th May 1976; to Nailstone Colliery, Leicestershire, 14th June 1976; to BR Doncaster, for repairs, 28th October 1976; to Nailstone Colliery, 23rd December 1976; to Great Central Railway, Loughborough, 6th February 1980.

D4068 Darlington 1961 40B 6/72 F No.56 / 9300-116

to NCB Ashington Central Workshops, January 1973; to Shilbottle Colliery, 16th February 1973; to Lambton Engine Works, Philadelphia, 3rd April 1979; to Whittle Colliery, Newton-on-the-Moor, 25th April 1980; scrapped on site by C.F. Booth Ltd of Rotherham, October 1985.

D4069 Darlington 1961 41J 4/72 F 9300-111

to NCB Ashington Central Workshops, September 1972; to Whittle Colliery, Newton-on-the-Moor, 20th October 1972; to Lambton Engine Works, Philadelphia, 28th February 1978; to Whittle Colliery, Newton-on-the-Moor, 30th March 1979; to C.F. Booth Ltd, Rotherham, November 1985; scrapped, December 1985.

D4070 Darlington 1961 41J 4/72 F No.52 / 9300-112

to NCB Ashington Central Workshops, September 1972; to Shilbottle Colliery, 9th February 1973; to Ashington Central Workshops, 23rd September 1974; to Lambton Engine Works, Philadelphia, for rebuild, 11th October 1975; to Bates Colliery, Blyth, 12th November 1976; to Whittle Colliery, Newton-on-the-Moor, 16th April 1977; to Lambton Engine Works, 29th April 1980; to Whittle Colliery, Newton-on-the-Moor, 5th June 1981; scrapped on site by C.F. Booth Ltd of Rotherham, October 1985.

D4072 Darlington 1961 31B 4/72 F No.53 / 9300-114

to NCB Ashington Central Workshops, October 1972; to Whittle Colliery, Newton-on-the-Moor, 1st November 1972; to Lambton Engine Works, Philadelphia, 14th June 1977; to Whittle Colliery, 28th February 1978; to Lambton Engine Works, 22nd June 1978; to Whittle Colliery, 16th August 1978; to Lambton Engine Works, 29th September 1978; to Whittle Colliery, 26th November 1978; to Lambton Engine Works, 19th December 1979; to Whittle Colliery, 5th June 1980; to Lambton Engine Works, 14th April 1981; to South Hetton Colliery, 10th May 1982; to Lambton Engine Works, 27th September 1982; to Ashington Colliery, July 1983; scrapped on site by C.F. Booth Ltd of Rotherham, November 1985.

D4074 Darlington 1961 31B 4/72 F No.54

to NCB Ashington Central Workshops, October 1972; to Whittle Colliery, Newton-on-the-Moor, 15th December 1972; to Lambton Engine Works, Philadelphia, 8th February 1977; scrapped, August 1978.

D4092 Darlington 1960 34E 9/68 P D4092

Powell Duffryn Fuels Ltd, NCBOE Gwaun-cae-Gurwen Disposal Point, Glamorgan, October 1968; to BR Canton Depot, Cardiff, for repairs, August 1977; returned to Gwaun-cae-Gurwen Disposal Point, October 1977; to South Yorkshire Railway Preservation Society (HNRC), Meadowhall, Sheffield, 26th October 1988; to Barrow Hill Engine Shed Society (HNRC), Staveley, 26th July 2001.

D4095 Horwich 1961 66B 2/04 P 881

08881 to Traditional Traction, Wishaw, Warwickshire, 20th April 2007; to Alstom Transport, Stonebridge Park Heavy Repair Depot, Wembley, for tyre turning, 15th August 2007; returned to Wishaw, 19th September 2007; to Gloucestershire Warwickshire Railway, Toddington, for storage, early 2008; to LaFarge Cement, Barrow-on-Soar, on hire, 5th October 2008; to Gloucestershire Warwickshire Railway, Toddington, December 2008; to Somerset & Dorset Railway Heritage Trust, Midsomer Norton, Somerset, 31st January 2012; to Gloucestershire Warwickshire Railway, Toddington, on loan, 5th December 2013; to Somerset & Dorset Railway Heritage Trust, Midsomer Norton, 17th March 2014.

D4115 Horwich 1962 36A 5/93 P 08885 / 18 / HO42

08885 to Great Central Railway, Ruddington, Nottingham, 15th June 1994; to Midland Railway, Butterley, 7th November 2004; purchased by RT Rail, Crewe, August 2005; to RMS Locotec, Dewsbury, for overhaul, 25th August 2005; to Network Rail, Whitemoor Yard, March, on hire, 31st October 2005; purchased by RMS Locotec, about 2007; to PD Ports, Tees Dock, on hire, 3rd April 2009; to RMS Locotec, Weardale Railway, Wolsingham, about 15th April 2013.

D4116 Horwich 1962 5A ? F 08886

08886 to Traditional Traction, Wishaw, Warwickshire; despatched from Crewe Electric Depot, 12th May 2016; used for spares; remains to European Metal Recycling, Kingsbury, for scrap, 20th May 2016; scrapped.

D4117 Horwich 1962 ? ? P 08887

08887 acquired by Alstom Transport, Longsight Depot, Manchester, 2007; to Alstom Transport, Wembley Depot, London, by 25th January 2012; to Arlington Fleet Services, Eastleigh Depot, 20th July 2016; to Alstom Transport, Wembley Depot, London, 19th December 2017; to Alstom Transport, Polmadie Depot, Glasgow, by 3rd May 2018.

D4118 Horwich 1962 ? ? P D4118

08888 to Kent & East Sussex Railway, Tenterden; despatched from DB Cargo, Hoo Junction, 15th December 2016.

D4121 Horwich 1962 ? ? P 08891

08891 to LH Group, Barton under Needwood, 3rd May 2008; to Nemesis Rail, Burton upon Trent, 16th August 2017.

D4122 Horwich 1962 70D 12/96 P 08892

08892 to RFS (Engineering) Ltd, Doncaster, 11th December 1996; to GNER, Bounds Green Depot, London, on hire, 18th April 1997; returned to RFS (Engineering) Ltd, Doncaster, 2nd August 1999; to GNER, Bounds Green Depot, London, on hire, by 21st April 2001; to Wabtec, Doncaster, 10th April 2003; to Soho EMU Depot, Birmingham, on hire, October 2004; to Central Trains, Tyseley, on hire, 12th November 2004; returned to Wabtec, Doncaster, June 2005; sold to Direct Rail Services, May 2006; to DRS, Kingmoor Depot, Carlisle, 25th July 2006; to DRS, Gresty Bridge Depot, Crewe, about January 2007; to North Pole Depot, near Kensal Green, London, on hire, October 2007; to DRS, Gresty Bridge Depot, Crewe, 14th November 2007; to Fastline, Doncaster, on hire, early July 2008; sold to Harry Needle Railroad Company, about August 2008; to LaFarge, Hope Cement Works, Derbyshire, on hire, by 6th September 2008; to Bombardier Transportation, Litchurch Lane Works, Derby, on hire, 14th October 2010; to Nemesis Rail, Burton upon Trent, 4th November 2011; to Bombardier Transportation, Litchurch Lane Works, Derby, on hire, 20th June 2013; to Barrow Hill Engine Shed Society, Staveley, 24th July 2013; to First Capital Connect, Hornsey Depot, London, on hire, 20th December 2013; to Serco, Old Dalby Test Centre, 19th February 2015; to Tyseley Steam Depot, Birmingham, for tyre turning, 29th July 2016; returned to Old Dalby Test Centre, 12th August 2016.

D4126 Horwich 1962 86A 2/04 P 08896 / STEVEN DENT

08896 to Traditional Traction, Wishaw, Warwickshire; despatched from EWS Toton Depot, 8th March 2007; to Severn Valley Railway, Bridgnorth, November 2009.

D4133 Horwich 1962 36A 7/95 P JOHN W. ANTILL

08903 to ICI Billingham Works, Stockton-on-Tees, 9th May 1996; to RFS (Engineering) Ltd, Doncaster, for repairs, March 1997; to ICI Wilton Works, Middlesbrough, 4th August 1997; to ICI Billingham Works, September 1999; to ICI Wilton Works, 2005; to ICI Billingham Works, 17th April 2007; to ICI Wilton Works, Middlesbrough, 2008; to Wensleydale Railway, Leeming Bar, for gala, 13th September 2016; to Sembcorp Utilities, Wilton, 27th September 2016.

D4134 Horwich 1962 70D ? P 08904

08904 sold to Harry Needle Railroad Company; despatched from Eastleigh, 23rd February 2017; to Celsa, Cardiff, on hire, 24th February 2017; to Harry Needle Railroad Company, Worksop Yard, 13th June 2019.

D4135 Horwich 1962 2F 1/05 P 08905

08905 sold to Harry Needle Railroad Company, 2011; to Hope Cement Works, Derbyshire, on hire, October 2011.

D4137 Horwich 1962 ? ? P 08907

08907 to Great Central Railway, Loughborough, 28th November 2016.

D4141 Horwich 1962 40B 2/04 P 08911 / MATEY

08911 to National Railway Museum, York; despatched from EWS Thornaby Depot, 15th May 2004; to Southall Railway Centre, London (for use in 'Railway Children' production), 23rd May 2010; to National Railway Museum, York, 8th January 2011; to Southall Railway Centre, London (for use in 'Railway Children' production), 26th May

2011; to National Railway Museum, York, 24th January 2012; to Freightliner, York, on hire, 28th January 2013; to National Railway Museum, York, March 2014; to National Railway Museum, Shildon, 9th June 2016.

D4142 Horwich 1962 8J 9/02 P 08912

08912 to T.J. Thomson Ltd, Stockton-on-Tees; despatched from EWS Toton Depot, 6th March 2007; to EWS Thornaby Depot, for tyre turning, 1st February 2008; to A.V. Dawson Ltd, Middlesbrough, 20th March 2008; used for spares, from May 2008; was dismantled, no wheels, off track, 30th March 2019.

D4143 Horwich 1962 66B 11/05 F 08913 / HYWELL

08913 to LH Group, Barton under Needwood; despatched from EWS Toton Depot, 20th March 2007; to Manchester Ship Canal Company, Barton Dock, on hire, 24th April 2007; purchased by Hunslet Engine Company, March 2008; to Cleveland Potash Ltd, Boulby Mine, on hire, May 2009; to LH Group, Barton under Needwood, 30th October 2009; to Daventry International Rail Freight Terminal, on hire, 10th December 2009; to LH Group, Barton under Needwood, for repairs, week-ending 24th June 2011; to Daventry International Rail Freight Terminal, on hire, July 2011; to LH Group, Barton under Needwood, 4th March 2013; to European Metal Recycling, Kingsbury, for scrap, 18th January 2018; scrapped.

D4145 Horwich 1962 8J 2/04 P 08915 / HERCULES

08915 to Traditional Traction, Wishaw, Warwickshire; despatched from EWS Toton Depot and moved direct to Colne Valley Railway, Castle Hedingham, 13th March 2007; to Stephenson Railway Museum, Chirton, near Newcastle upon Tyne, 5th November 2009.

D4148 Horwich 1962 TM 7/05 P 08918

08918 sold to Harry Needle Railroad Company; to Nemesis Rail, Burton upon Trent, October 2011.

D4151 Horwich 1962 ? ? P 08921 / PONGO

08921 to European Metal Recycling, Kingsbury, 5th August 2011; to Railway Support Services, Wishaw, Warwickshire, 1st February 2018.

D4152 Horwich 1962 16A ? P 08922

08922 to Great Central Railway, Ruddington, Nottingham; despatched from DB Cargo, Toton Depot, 14th December 2016; to DB Cargo Maintenance Ltd, Wheildon Road Wagon Works, Stoke on Trent, on hire, 28th March 2019.

D4154 Horwich 1962 66B 8/06 P 08924 / CELSA 2

08924 to C.F. Booth Ltd, Rotherham; despatched from DB Schenker, Tyne Yard, 27th January 2011; purchased by Harry Needle Railroad Company; to Barrow Hill Engine Shed Society, Staveley, 15th February 2011; to LaFarge, Hope Cement Works, Derbyshire, on hire, 2nd April 2013; to Barrow Hill Engine Shed Society, Staveley, 12th February 2014; to GBRf, Garston Railport, Liverpool, on hire, 21st February 2014; to Barrow Hill Engine Shed Society, Staveley, 21st December 2015; to Celsa, Cardiff, on hire, 15th March 2016; to Barrow Hill Engine Shed Society, Staveley, 3rd October 2016; to Tyseley Steam Depot, Birmingham, for tyre turning, 12th October 2016; to Barrow Hill Engine Shed Society,

Staveley, 17th October 2016; to Celsa, Cardiff, on hire, 25th November 2016; to Barrow Hill Engine Shed Society, Staveley, 13th February 2017; to Celsa, Cardiff, on hire, 3rd March 2017; to Barrow Hill Engine Shed Society, Staveley, 31st March 2017; to Celsa, Cardiff, on hire, 7th April 2017; to Barrow Hill Engine Shed Society, Staveley, for repairs, 10th June 2019.

D4155 Horwich 1962 ? ? P 08925
08925 purchased by GBRf, but to be maintained by Harry Needle Railroad Company; to Barrow Hill Engine Shed Society, Staveley, 23rd August 2016; to Railway Support Services, Wishaw, Warwickshire, 18th November 2016; to Midland Road Depot, Leeds, for tyre turning, 20th November 2016; to GBRf, Immingham, November 2016; to Barrow Hill Engine Shed Society, Staveley, December 2016; to GBRf, Whitemoor Yard, March, 3rd February 2017; to Barrow Hill Engine Shed Society, Staveley, for repairs, May 2018; to GBRf, Whitemoor Yard, March, 13th June 2018.

D4156 Horwich 1962 8J 11/96 F 08926
08926 to Traditional Traction, Wishaw, Warwickshire, 28th March 2007; used for spares; remains to European Metal Recycling, Kingsbury, for scrap, 17th June 2007; scrapped, July 2007.

D4157 Horwich 1962 66B 6/05 P D4157 / 08927
08927 to Traditional Traction, Wishaw, Warwickshire, and moved direct to Gloucestershire Warwickshire Railway, Toddington, 28th March 2007; to Alstom Transport, Stonebridge Park Heavy Repair Depot, Wembley, for tyre turning, 19th September 2007; returned to Traditional Traction (GWR), 24th November 2007; to Pontypool & Blaenavon Railway, on hire, 28th April 2010; to Gloucestershire Warwickshire Railway, Toddington, 31st January 2011; to Southall Railway Centre, London (for use in Railway Children production), May 2011; to Gloucestershire Warwickshire Railway, Toddington, 31st January 2012; to National Railway Museum, Shildon, on hire, 16th July 2012; to Electro-Motive Diesel Ltd, Roberts Road Depot, Doncaster, on hire, 5th November 2013; to Railway Support Services, Wishaw, 8th October 2018; to DB Cargo Maintenance Ltd, Whcidon Road Wagon Works, Stoke on Trent, on hire, 15th December 2018; to Bescot Yard, Walsall, on hire, 28th March 2019.

D4158 Darlington 1962 NC ? F 08928
08928 sold to Cotswold Rail, Moreton in Marsh, about June 2001; to Mid-Norfolk Railway, Dereham, for loading; despatched by rail from Crown Point Depot, Norwich, 14th August 2001; to Brush, Loughborough, for repairs, 15th August 2001; sold to Harry Needle Railroad Company; to Barrow Hill Engine Shed Society, Staveley, 17th July 2003; to HNRC, Long Marston, for storage, 21st March 2006; to European Metal Recycling, Kingsbury, for scrap, 3rd December 2010; scrapped, December 2010.**D4163**

D4163 Darlington 1962 81A 2/08 P 08933
08933 sold to T.J. Thomson Ltd, Stockton-on-Tees, early 2009; resold direct to Foster Yeoman Quarries Ltd, Merehead Stone Terminal, Somerset; despatched from Hoo Junction, 13th March 2009; to Knight's Rail Services, Eastleigh Works, for repairs, 19th October 2010; to Merehead Stone Terminal, Somerset, 3rd July 2012.

D4166 Darlington 1962 31B 12/92 P 08936 / HO75
08936 to South Yorkshire Railway Preservation Society (HNRC), Meadowhall,

Sheffield, 31st January 1994; to Railway Age, Crewe, on hire, 22nd January 1999; to Fragonset, Derby, for overhaul, 9th June 2000; to Barrow Hill Engine Shed Society (HNRC), Staveley, 4th December 2001; to Fragonset, Derby, May 2002; to Barrow Hill Engine Shed Society, Staveley, 4th February 2003; to Network Rail, Whitemoor Yard, March, on hire, about May 2004; purchased by Cotswold Rail, Moreton in Marsh, about July 2004; to Allelys Ltd, Studley, Warwickshire, for storage, by 31st January 2006; to Alstom Transport, Stonebridge Park Heavy Repair Depot, Wembley, on hire, by 23rd February 2006; to Willesden Depot, 28th May 2006; to Horton Road Depot, Gloucester, for storage, early November 2006; purchased by RMS Locotec, 2007; to RMS Locotec, Wakefield, for repairs, 4th October 2007; to Corus, Shotton, on hire, November 2007; to RMS Locotec, Weardale Railway, Wolsingham, 17th August 2016.

D4167 Darlington 1962 84A 12/93 P D4167
08937 to Aggregate Industries Ltd, Meldon Quarry, Devon, 1995; used at the quarry and by the Dartmoor Railway; to RMS Locotec, Dewsbury, for repairs, 25th November 2005; to Wabtec, Doncaster, for sub-contract repairs, May 2006; to Aggregate Industries, Meldon Quarry, Devon, 24th April 2007; quarry mothballed, July 2011; locomotive continued to be used by Dartmoor Railway.

D4169 Darlington 1962 16A ? P 08939
08939 to Railway Support Services, Wishaw, Warwickshire; despatched from DB Schenker, Toton Depot, August 2015; to Colne Valley Railway, Castle Hedingham, for storage, 22nd September 2015; to Railway Support Services, Wishaw, 21st March 2017; to East Midlands Trains, Neville Hill Depot, Leeds, on hire, 29th May 2019; to GBRf, Port of Felixstowe, on hire, 12th July 2019.

D4173 Darlington 1962 1A 7/88 P 08943
08943 to ABB Transportation, Crewe, April 1989; locomotive included in sale when BREL works privatised; to ABB Transportation, Derby, September 1989; to ABB Transportation, York, February 1993; to ABB British Wheelset Ltd, Trafford Park, Manchester, about August 1996; to ABB Transportation, Crewe, May 1998; purchased by Harry Needle Railroad Company, July 2009; to Barrow Hill Engine Shed Society, Staveley, 12th February 2010; to Southall Railway Centre, London (for use in connection with Railway Children play at Waterloo Station), on hire, 25th May 2010; to Bombardier Transportation, Central Rivers Depot, Barton under Needwood, on hire, 7th January 2011.

D4174 Darlington 1962 81A 5/98 P 08944
08944 to Mike Darnall, Newton Heath, Manchester, November 2000; to Wabtec, Doncaster, for overhaul, about December 2000; to East Lancashire Railway, Bury, by April 2001; to Crewe Electric Depot, for tyre turning, 7th March 2007; to East Lancashire Railway, Bury, 15th March 2007; purchased by Harry Needle Railroad Company, and stored at Bury, August 2015.

D4176 Darlington 1962 8J 2/02 F 08946
08946 to Traditional Traction, Wishaw, Warwickshire, 23rd March 2007; scrapped by Moveright International, Wishaw, June 2008.

D4177 Darlington 1962 81A 3/04 P 08947 / MARGARET
08947 sold to T.J. Thomson Ltd, Stockton-on-Tees, 2007; resold direct to Foster

Yeoman Quarries Ltd, Merehead Stone Terminal; despatched from EWS Westbury, 13th April 2007; to Whatley Quarry, by 9th September 2007; to Merehead Stone Terminal, initially for open day held on 22nd June 2008; to Isle of Grain Stone Terminal, 6th June 2012; to Arlington Fleet Services, Eastleigh Works, for overhaul, 3rd May 2013; to Whatley Quarry, 2013; to Arlington Fleet Services, Eastleigh Works, for repairs, by 3rd October 2015; to Whatley Quarry, 23rd December 2015.

D4178 Darlington 1962 ? ? P 08948

08948 to Eurostar UK Ltd at privatisation, 1994; fitted with a Sharfenberg coupler; based at Eurostar, North Pole Depot, London; to Barrow Hill Engine Shed Society, Staveley, for repairs, 17th September 2004; to Eurostar, North Pole Depot, London, 21st February 2006; to Eurostar, Temple Mills Depot, Leyton, London, October 2007.

D4183 Darlington 1963 16A 1/06 F 08953

08953 to European Metal Recycling, Attercliffe, Sheffield; despatched from DB Schenker, Doncaster Depot, 16th May 2010; scrapped, early May 2012.

D4184 Darlington 1963 2F ? P 08954

08954 sold to Harry Needle Railroad Company; to Boden Rail Engineering, Washwood Heath, for repairs; despatched from DB Schenker, Toton Depot, 15th February 2011; to Nemesis Rail, Burton upon Trent, 18th October 2011; purchased by Alstom Transport about May 2013; to Alstom Transport, Longsight Depot, Manchester, 29th May 2013, to Alstom Transport, Polmadie Depot, Glasgow, 1st May 2014; to Arlington Fleet Services, Eastleigh Works, for repairs, 10th August 2016; to Alstom Transport, Edge Hill Depot, Liverpool, 21st September 2016; to Alstom Transport, Wembley Depot, London, 15th December 2016; to Alstom Transport, Polmadie Depot, Glasgow, 25th January 2017.

D4186 Darlington 1963 ? ? P 08956

08956 to Serco Railtest, Derby, about August 2001; to Fragonset Rail, Derby, by May 2005; to Serco Railtest, Derby, about July 2007; to Metronet Rail, Old Dalby Test Centre, 9th December 2008; to Moveright International, Wishaw, Warwickshire, for repairs, 19th February 2016; returned to Old Dalby Test Centre, June 2016.

SECTION 17

British Railways built 0-6-0 diesel electric locomotives, numbered D3665-D3671, D3719-D3721, and D4099-D4114. Basically the same as a 08 shunter, but capable of a higher speed (27.5mph instead of 20mph). Later classified TOPS Class 09.

D3665 Darlington 1959 16A 2/09 P 09001

09001 to Peak Rail, Rowsley; despatched from DB Schenker, Doncaster Depot, 28th January 2011.

D3666 Darlington 1959 75C 9/92 P D3666

09002 to South Devon Railway, Buckfastleigh, 11th June 1993; sold to GBRf, early 2011; to LH Group, Barton under Needwood, for repairs, 2nd March 2011; to Great Central Railway, Ruddington, Nottingham, for running-in, 29th September 2011; to

Freightliner, Trafford Park, Manchester, on hire, 1st October 2011; to GBRf, Whitemoor Yard, March, January 2013; to Moveright International, Wishaw, Warwickshire, 3rd February 2017; to Barrow Hill Engine Shed Society, Staveley, 6th February 2017.

D3667 Darlington 1959 81A 5/08 F 09003
09003 sold to Harry Needle Railroad Company, July 2010; to Boden Rail Engineering, Washwood Heath; despatched from Margam Depot; used for spares, August 2010; remains to European Metal Recycling, Kingsbury, for scrap, 19th September 2011; scrapped September 2011.

D3668 Darlington 1959 75C 4/99 P D3668
09004 to Lavender Line, Isfield, 14th December 2000; to Spa Valley Railway, Tunbridge Wells, 12th March 2003; to Swindon & Cricklade Railway, 27th June 2009; to Avon Valley Railway, Bitton, 10th April 2014; to St Philip's Marsh Depot, Bristol, for tyre turning, 29th May 2014; to Swindon & Cricklade Railway, 5th June 2014.

D3670 Darlington 1959 16A ? P 09006
09006 sold to Harry Needle Railroad Company; to Nemesis Rail, Burton upon Trent; despatched from DB Schenker, Toton Depot, 15th September 2015.

D3671 Darlington 1959 16A 12/09 P D3671
09007 to London Overground, Willesden Depot, London, 23rd September 2010.

D3719 Darlington 1959 87B 5/04 F 09008
09008 to Harry Needle Railroad Company; to Boden Rail Engineering, Washwood Heath; despatched from Bescot Depot, 7th January 2011; used for spares; remains to European Metal Recycling, Kingsbury, for scrap, 19th September 2011; scrapped September 2011.

D3720 Darlington 1959 81A 5/04 P 09009
09009 to C.F. Booth Ltd, Rotherham; despatched from DB Schenker, Toton Depot, 25th January 2011; resold to GBRf; to LH Group, Barton under Needwood, for repairs, 14th February 2011.

D3721 Darlington 1959 81A 9/04 P D3721
09010 to South Devon Railway, Buckfastleigh; despatched from DB Schenker, Hither Green Depot, 30th September 2010; to St Philip's Marsh Depot, Bristol, for tyre turning, 30th August 2012; to South Devon Railway, Buckfastleigh, 5th September 2012.

D4100 Horwich 1961 81A 2/04 P D4100 / DICK HARDY
09012 to Barrow Hill Engine Shed Society (HNRC), Staveley; despatched from DB Schenker, Hither Green Depot, September 2010; to Severn Valley Railway, Bridgnorth, on hire, 21st February 2013.

D4102 Horwich 1961 16A 3/09 P 09014
09014 sold to Harry Needle Railroad Company; to Boden Rail Engineering, Washwood Heath, for repairs; despatched from DB Schenker, Doncaster Depot, 21st January 2011; to Nemesis Rail, Burton upon Trent, 13th September 2012.

D4103 Horwich 1961 87B 3/07 P 09015 / ROB

09015 to T.J. Thomson Ltd, Stockton-on-Tees, 21st February 2011; purchased by National Railway Museum, York, but moved direct to Moveright International, Wishaw, Warwickshire, for storage, 29th March 2011; sold to Railway Support Services, Wishaw, 1st February 2017; used for spares; remains to Avon Valley Railway, Bitton, 10th July 2019.

D4105 Horwich 1961 16A 3/10 P 09017 / LEO

09017 purchased by National Railway Museum, York; despatched from DB Schenker, Toton Depot; stored at Network Rail, Klondyke Yard, York, 10th August 2011; to Freightliner Maintenance Ltd, York, on hire, March 2014; to National Railway Museum, York, by June 2016.

D4106 Horwich 1961 81A 7/04 P D4106

09018 sold to Harry Needle Railroad Company; to Boden Rail Engineering, Washwood Heath, for repairs; despatched from DB Schenker, Hither Green Depot, 29th September 2010; to Metronet Rail, Old Dalby Test Centre, Leicestershire, on hire, 19th October 2011; to LaFarge, Hope Cement Works, Derbyshire, on hire, 7th June 2012; to Bluebell Railway, Horsted Keynes, on hire, 23rd April 2013; to Bombardier Transportation, Ilford Depot, for tyre turning, 9th December 2015; to Bluebell Railway, Horsted Keynes, on hire, 10th December 2015.

D4107 Horwich 1961 16A 6/10 P D4107

09019 sold to Harry Needle Railroad Company; to Barrow Hill Engine Shed Society, Staveley; despatched from DB Schenker, Toton Depot, 16th February 2011; to Nemesis Rail, Burton upon Trent, 10th February 2012; to West Somerset Railway, Minehead, on hire, 5th March 2013.

D4110 Horwich 1962 16A 2/11 P 09022

09022 sold to Harry Needle Railroad Company; to Boden Rail Engineering, Washwood Heath, for repairs, 8th August 2011; resold to Port of Boston; to Port of Boston, 16th April 2012; to Tyseley Steam Depot, Birmingham, for repairs, 8th February 2016; to Port of Boston, 16th February 2016.

D4111 Horwich 1962 40B ? P 09023

09023 to European Metal Recycling, Kingsbury; despatched from DB Schenker, Immingham Depot, 24th August 2011; to European Metal Recycling Ltd, Attercliffe, Sheffield, 18th September 2014.

D4112 Horwich 1962 81A 5/08 P 09024

09024 to C.F. Booth Ltd, Rotherham; despatched from Eastleigh, 26th October 2011; to Goodman, Wishaw, Warwickshire, for storage, 10th December 2011; to East Lancashire Railway, Bury, 23rd January 2014.

D4113 Horwich 1962 75C 10/05 P 09025 / D4113

09025 to East Kent Railway, Shepherdswell; despatched from Selhurst Depot, about 25th September 2005; to Lavender Line, Isfield, 10th October 2014.

D4114 Horwich 1962 ? ? P 09026

09026 to Spa Valley Railway, Tunbridge Wells; despatched from Brighton, 22nd May 2016.

D3927 Horwich 1961 55G 12/16 P 09106 / 6

09106 to Harry Needle Railroad Company; despatched from DB Schenker, Knottingley Depot; former number 08759; to DRC, Ferrybridge, for storage, March 2017; to Barrow Hill Engine Shed Society, Staveley, 10th July 2017; to GBRf, Dagenham, on hire, 20th October 2017; to Barrow Hill Engine Shed Society, Staveley, 7th December 2017; to Railway Support Services, Wishaw, Warwickshire, 7th June 2019; to Celsa, Cardiff, on hire, 10th June 2019.

D4013 Horwich 1961 ? ? P 09107

09107 to European Metal Recycling, Kingsbury, September 2011; former number 08845; to Severn Valley Railway, Bridgnorth, 15th June 2017.

D3536 Derby 1958 16A ? P 09201

09201 sold to Harry Needle Railroad Company; despatched from DB Schenker, Toton Depot; former number 08421; to Hope Cement Works, Derbyshire, on hire, 14th September 2015.

D3884 Crewe 1960 40B 2/04 P 09204

09204 to London & North Western Railway Company Ltd, Tyne Yard Depot, Lamesley, Gateshead (ex DB Schenker, with site), 21st April 2011; to L&NWR, Carriage Works, Crewe, 3rd August 2012; to Crewe Electric Depot, for tyre turning, by 13th August 2012; to L&NWR, Carriage Works, Crewe, 29th August 2012.

SECTION 18

Clayton Equipment Co Ltd built Bo-Bo diesel electric locomotives, numbered D8500-D8616, and introduced 1962. Fitted with two Paxman 6ZHXL engines (each developing 450bhp at 1500rpm) and driving wheels of 3ft 3½in diameter. Later classified TOPS Class 17, with three sub-divisions of which D8568 was 17/1.

D8568 CE 4365U/69 1963 66A 10/71 P D8568

to Hemel Hempstead Lightweight Concrete Co Ltd, Cupid Green, Hertfordshire, 11th September 1972; to Ribblesdale Cement Ltd, Clitheroe, Lancashire, 16th June 1977; arrived at Clitheroe, 24th June 1977; sold to North Yorkshire Moors Railway, Grosmont, December 1982; left Clitheroe, 9th February 1983; arrived at North Yorkshire Moors Railway, Grosmont, 11th February 1983; to BR Gloucester, 2nd August 1991; exhibited at BR Gloucester Depot open day, 18th August 1991; to BR Old Oak Common Depot, London, open day, August 1991; to BR Stonebridge Park Sidings, London, November 1991; to BR Willesden Depot, for tyre turning, December 1991; to Chinnor and Princes Risborough Railway, Oxfordshire, 25th April 1992; to Severn Valley Railway, Bridgnorth, for gala, October 1998; returned to Chinnor and Princes Risborough Railway; to Old Oak Common Depot, London, open day, August 2000; returned to Chinnor and Princes Risborough Railway; to Barrow Hill Engine Shed Society, Staveley, for open day, October 2001; returned to Chinnor and Princes Risborough Railway, 9th October 2001; to Severn

Valley Railway, Bridgnorth, for gala, 8th September 2015; returned to Chinnor and Princes Risborough Railway, 13th October 2015; to Severn Valley Railway, Bridgnorth, for gala, 16th May 2017; returned to Chinnor and Princes Risborough Railway; to South Devon Railway, Buckfastleigh, for gala, early November 2017; returned to Chinnor and Princes Risborough Railway, 7th November 2017.

SECTION 19

British Railways built 0-6-0 diesel hydraulic locomotives, numbered D9500-D9555, and introduced 1964. Fitted with a Paxman 'Ventura' 6YJX engine developing 650bhp at 1500rpm, and driving wheels of 4ft 0in diameter. Later classified TOPS Class 14.

D9500 Swindon 1964 86A 5/69 P D9500

to NCB Ashington Colliery, November 1969; despatched from Canton Depot, Cardiff, 17th November 1969; to Lambton Engine Works, Philadelphia, 12th July 1978; to Ashington Colliery, 26th June 1979; to BR Thornaby Depot, for repairs, by 14th May 1982; returned to Ashington Colliery, by 24th May 1982; to Lambton Engine Works, 12th May 1983; to Ashington Colliery, 25th July 1983; to Llangollen Railway, 25th September 1987; to Heritage Centre, Swindon, April 1988; to West Somerset Railway, Minehead, 4th November 1989; to South Yorkshire Railway Preservation Society, Meadowhall, Sheffield, August 1992; to Barrow Hill Engine Shed Society, Staveley, 2nd August 2001; to Peak Rail, Rowsley, 13th May 2008; to Andrew Briddon, Darley Dale, 29th May 2015.

D9502 Swindon 1964 86A 5/69 P D9502

to NCB Ashington Colliery; despatched from Canton Depot, Cardiff, 30th June 1969; arrived at Ashington, early July 1969; to Burradon Colliery, Dudley, by September 1969; to Ashington Central Workshops, about July 1973; to Burradon Colliery, Dudley, March 1974; to Backworth Colliery, January 1976; to Weetslade Coal Preparation Plant, about June 1976; to Ashington Colliery, 24th April 1981; to Lambton Engine Works, 28th April 1983; to Ashington Colliery, 12th May 1983; to Llangollen Railway, 25th September 1987; to South Yorkshire Railway Preservation Society (HNRC), Meadowhall, Sheffield, 12th March 1992; to Peak Rail, Rowsley, March 2002; to East Lancashire Railway, Bury, 20th November 2014.

D9503 Swindon 1964 50B 4/68 F 65 / 8411-25

to Stewarts & Lloyds Minerals Ltd, Harlaxton Quarries, Lincolnshire, November 1968; to Corby Quarries, July 1974; scrapped, September 1980.

D9504 Swindon 1964 50B 4/68 P D9504

to NCB Philadelphia Locomotive Shed, County Durham; despatched from 50B Dairycoates Depot, Hull, 27th November 1968; arrived at Philadelphia, 29th November 1968; to Boldon Colliery, 21st August 1973; to Lambton Engine Works, Philadelphia, 7th September 1973; to Boldon Colliery, February 1974; to Backworth Colliery, 17th December 1974; seen at BR Cambois Depot, 3rd January 1975; to Burradon Colliery, Dudley, 29th January 1975; to Weetslade Coal Preparation Plant, 3rd January 1976; to Lambton Engine Works, Philadelphia, 21st April 1981; to Ashington Colliery, 11th September 1981; sold to Kent & East Sussex Railway, Tenterden, 26th September 1987;

to Nene Valley Railway, Wansford, for overhaul, 25th February 1998; returned to Kent & East Sussex Railway, Tenterden, 21st April 1999; to Channel Tunnel Rail Link, Beechbrook Farm, near Ashford, on hire, 1st August 2001; to Medway Ports, Chatham Dockyard, on hire, February 2003; to Nene Valley Railway, Wansford, for repairs, 4th April 2003; to EWS Toton Depot, for tyre turning, 7th January 2004; returned to Nene Valley Railway, Wansford, 9th January 2004; to Victa Rail, March, on hire, 15th January 2004; to Nene Valley Railway, Wansford, for repairs, 8th April 2004; to Channel Tunnel Rail Link, Swanscombe, on hire, 15th June 2004; to Channel Tunnel Rail Link, Dagenham, on hire, 2nd November 2004; to Nene Valley Railway, Wansford, for repairs, 24th February 2005; to Channel Tunnel Rail Link, Dagenham, on hire, by 3rd May 2005; to Nene Valley Railway, Wansford; to Channel Tunnel Rail Link, Swanscombe, on hire, 7th November 2005; to Nene Valley Railway, Wansford, 13th January 2006; to Channel Tunnel Rail Link, Dagenham, on hire, 2nd March 2006; to Nene Valley Railway, Wansford, 25th January 2007; to Aggregate Industries Ltd, Bardon Hill, on hire, 16th January 2008; to Nene Valley Railway, Wansford, 10th February 2009; to Kent & East Sussex Railway, Tenterden, 5th May 2010.

D9505 Swindon 1964 50B 4/68 F MICHLOW

to APCM, Hope Cement Works, Derbyshire, 26th September 1968; sold for export (see Appendix C); left Hope on 5th May 1975.

D9507 Swindon 1964 50B 4/68 F 55 / 8311-35

to Stewarts & Lloyds Minerals Ltd, Corby Quarries, November 1968; to BSC Steelworks Disposal Site, Corby, December 1980; scrapped on site by Shanks & McEwan Ltd of Corby, September 1982.

D9508 Swindon 1964 87E 10/68 F No.9 / 9312-99

to NCB Ashington Colliery; despatched from BR Canton Depot, Cardiff, 6th March 1969; withdrawn November 1983; scrapped on site by D. Short Ltd of North Shields, 17th January 1984.

D9510 Swindon 1964 50B 4/68 F 60 / 8411-23

to Stewarts & Lloyds Minerals Ltd, Buckminster Quarries, Lincolnshire, December 1968; seen at BR Grantham Station, 31st August 1972; to Corby Quarries, 6th September 1972; to BSC Tube Works, Corby, January 1981; scrapped on site by Shanks & McEwan Ltd of Corby, August 1982.

D9511 Swindon 1964 50B 4/68 F 9312-98

to NCB Ashington Colliery, January 1969; to Bates Colliery, Blyth, April 1969; to Burradon Colliery, Dudley, by 25th May 1969; to Ashington Colliery, about October 1972; dismantled for spares after a fire; remains scrapped, July 1979.

D9512 Swindon 1964 50B 4/68 F 63 / 8411-24

to Stewarts & Lloyds Minerals Ltd, Buckminster Quarries, Lincolnshire, December 1968; seen at BR Grantham Station, 31st August 1972; to Corby Quarries, 6th September 1972; used for spares; to BSC Steelworks Disposal Site, Corby, 29th December 1980; scrapped, about February 1982.

D9513 Swindon 1964 86A 3/68 P NCB 38

to W.H. Arnott Young & Co Ltd, Parkgate, Rotherham; despatched from 85A Worcester Depot, 18th July 1968; re-sold to Hargreaves Industrial Services Ltd, NCBOE British Oak Disposal Point, Crigglestone, November 1968; to NCBOE Bowers Row Disposal Point, Astley, 5th September 1969; sold to NCB North East Area; to Allerton Bywater Central Workshops, West Yorkshire, for overhaul, October 1973; to Ashington Colliery, January 1974; to Backworth Colliery, July 1974; to Burradon Colliery, Dudley, July 1974; to Backworth Colliery, 5th January 1976; to Lambton Engine Works, Philadelphia, 22nd November 1976; to Ashington Colliery, 14th February 1977; to Lambton Engine Works, 8th September 1977; to Ashington Colliery, 14th November 1977; to Lambton Engine Works, 30th June 1982; to Ashington Colliery, 17th February 1983; sold to C.F. Booth Ltd of Rotherham, summer 1987; did not leave Ashington site and was re-sold by Booth's for preservation; to Embsay & Bolton Abbey Railway, 12th October 1987; to East Lancashire Railway, Bury, for gala, 22nd July 2014; to Embsay & Bolton Abbey Railway, 1st August 2014.

D9514 Swindon 1964 86A 5/69 F No.4 / 9312-96

to NCB Ashington Colliery; despatched from Canton Depot, Cardiff, 30th June 1969; arrived at Ashington, early July 1969; to BR Gosforth Depot, for repairs, 11th October 1975; to Ashington Colliery, by 21st November 1975; to Lambton Engine Works, Philadelphia, about August 1977; to Ashington Colliery, about 1978; to Lambton Engine Works, September 1980; returned to Ashington Colliery; to BR Thornaby Depot, for repairs, October 1982; returned to Ashington Colliery; scrapped on site, November 1985.

D9515 Swindon 1964 50B 4/68 F 62 / 8411-22

to Stewarts & Lloyds Minerals Ltd, Buckminster Quarries, Lincolnshire, 2nd November 1968; seen at BR Grantham Station, 31st August 1972; to Corby Quarries, 6th September 1972; to BSC Steelworks Disposal Site, Corby, 29th December 1980; to Hunslet Engine Co Ltd, Leeds, December 1981; overhauled and converted to 5ft 6in gauge; exported to Bilbao, Spain from Goole Docks, circa 16th June 1982; scrapped, by 2002.

D9516 Swindon 1964 50B 4/68 P D9516

to Stewarts & Lloyds Minerals Ltd, Corby Quarries, November 1968; to BSC Steelworks Disposal Site, Corby, 29th December 1980; to Great Central Railway, Loughborough, 17th October 1981; to Severn Valley Railway, Bridgnorth, for diesel weekend, 15th October 1988; returned to Great Central Railway; to Nene Valley Railway, Wansford, 8th December 1988; to Boden Rail Engineering, Washwood Heath, 1st April 2011; to Wensleydale Railway, Leeming Bar, 11th April 2011; to Midland Road Depot, Leeds, for tyre turning, 21st February 2014; to Nene Valley Railway, Wansford, for repairs, 14th March 2014; to Didcot Railway Centre, Oxfordshire, 14th May 2014; to Old Oak Common Depot, London, for open day, September 2017; returned to Didcot Railway Centre, Oxfordshire, September 2017.

D9517 Swindon 1964 86A 10/68 F No.8 / 9312-93

to NCB Ashington Colliery, November 1969; despatched from BR Canton Depot, Cardiff, 17th November 1969; to Lambton Engine Works, Philadelphia, 14th June 1977; to Ashington Colliery, 5th September 1977; withdrawn November 1983; scrapped on site by D. Short Ltd of North Shields, January 1984.

D9518 Swindon 1964 86A 5/69 P No.7 / 9312-95

to NCB Ashington Colliery; despatched from Canton Depot, Cardiff, 30th June 1969; arrived at Ashington, early July 1969; to Lambton Engine Works, Philadelphia, June 1975; to Ashington Colliery, September 1975; to Lambton Engine Works, 5th September 1980; to Ashington Colliery, 3rd December 1980; to Rutland Railway Museum, Cottesmore, 26th September 1987; to Nene Valley Railway, Wansford, 8th September 2006; to West Somerset Railway, Minehead, 1st December 2011.

D9520 Swindon 1964 50B 4/68 P D9520 / 45

to Stewarts & Lloyds Minerals Ltd, Glendon Quarries, Northamptonshire, 16th December 1968; to Corby Quarries, 12th January 1970; to BSC Tube Works, Corby, October 1980; to North Yorkshire Moors Railway, Grosmont, 16th March 1981; to Rutland Railway Museum, Cottesmore, 21st February 1984; to Great Central Railway, Loughborough, on loan, 5th October 1985; returned to Cottesmore, 2nd December 1985; to Great Central Railway, Ruddington, Nottingham, 6th March 1998; to Nene Valley Railway, Wansford, 21st April 2004; to West Somerset Railway, Minehead, for gala, 15th June 2007; returned to Nene Valley Railway, Wansford; to Appleby Frodingham RPS, Scunthorpe, for gala, 9th to 11th May 2008; to National Railway Museum, York, May 2008; to Barrow Hill Engine Shed Society, Staveley, for gala, August 2008; to LaFarge, Hope Cement Works, Derbyshire, for open day, 6th September 2008; to Nene Valley Railway, Wansford, 2008; to West Somerset Railway, Minehead, for gala, June 2009; to Nene Valley Railway, Wansford, 14th June 2009; to West Somerset Railway, Minehead, for gala, 7th June 2010; to Nene Valley Railway, Wansford, 14th June 2010; to East Lancashire Railway, Bury, for gala, 23rd July 2014; to Nene Valley Railway, Wansford, 29th July 2014.

D9521 Swindon 1964 87E 5/69 P D9521

to NCB Ashington Colliery, 6th March 1970; to BR Gosforth Depot, Newcastle upon Tyne, for repairs, 11th March 1976; to Ashington Colliery, 17th April 1976; to Lambton Engine Works, Philadelphia, 30th November 1977; to Ashington Colliery, by October 1978; to Lambton Engine Works, 7th January 1982; to Ashington Colliery, 30th June 1982; to Lambton Engine Works, March 1983; to Ashington Colliery, March 1983; sold to C.F. Booth Ltd of Rotherham, summer 1987; did not leave Ashington site and was re-sold by Booth's for preservation; to Rutland Railway Museum, Cottesmore, 14th October 1987; to Swanage Railway, Dorset, 29th January 1992; to Wimbledon Depot, for tyre turning, 8th June 1995; returned to Swanage Railway, 12th June 1995; to Barry Rail Centre, Barry Island, 11th November 2004; to Mid-Norfolk Railway, Dereham, 20th May 2008; to Quainton Railway Society, near Aylesbury, Buckinghamshire, 9th July 2008; to Barry Rail Centre, Barry Island, autumn 2008; to Dean Forest Railway, Lydney, 16th January 2009; to Swindon & Cricklade Railway, for gala, September 2009; returned to Dean Forest Railway, Lydney, October 2009; to Gwili Railway, Bronwydd Arms, on loan, 10th August 2010; to Llangollen Railway, on loan, 25th August 2010; returned to Dean Forest Railway, Lydney, by 24th April 2011; to Avon Valley Railway, Bitton, 6th April 2013; to Dean Forest Railway, Lydney, May 2013; to East Lancashire Railway, Bury, for gala, 8th July 2014; to Dean Forest Railway, Lydney, 5th August 2014.

D9523 Swindon 1964 50B 4/68 P D9523

to Stewarts & Lloyds Minerals Ltd, Glendon East Quarries, Northamptonshire, 16th December 1968; to Corby Quarries, 28th May 1980; to BSC Steelworks Disposal Site,

29th December 1980; to Great Central Railway, Loughborough, 17th October 1981; to Nene Valley Railway, Wansford, 7th December 1988; to Boden Rail Engineering, Washwood Heath, 8th April 2011; to Derwent Valley Light Railway, Murton, York, 21st April 2011; to Cholsey & Wallingford Railway, Oxfordshire, for gala, 25th April 2013; to Derwent Valley Light Railway, Layerthorpe, York, 17th May 2013; to Wensleydale Railway, Leeming Bar, 18th July 2013; to East Lancashire Railway, Bury, for gala, 15th July 2014; to Nene Valley Railway, Wansford, for repairs, 13th August 2014; to Wensleydale Railway, Leeming Bar, 2nd February 2017.

D9524 Swindon 1964 87E 5/69 P 14901

to BP Refinery Ltd, Grangemouth; despatched from Canton Depot, Cardiff, July 1970; fitted with a Dorman type 8QT 500hp engine, by Andrew Barclay, Sons & Co Ltd; to BR Grangemouth Depot, for repairs, November 1971; returned to BP Refinery Ltd; to BR Eastfield Depot, for repairs, March 1978; returned to BP Refinery Ltd; to Scottish Railway Preservation Society, Falkirk, 9th September 1981; to Scottish Railway Preservation Society, Bo'ness, 7th February 1988; fitted with Rolls-Royce type DV8 750hp engine; sold to Middle Peak Railways, about June 2006; to RMS Locotec, Wakefield, for overhaul, 1st July 2006; to Elsecar Steam Railway, near Barnsley, 17th April 2007; to Midland Railway, Butterley, for gala, 18th May 2010; to Peak Rail, Rowsley, 1st July 2010; to Gwili Railway, Bronwydd Arms, on hire, 2nd April 2011; to Peak Rail, Rowsley, 22nd March 2013; to East Lancashire Railway, Bury, for gala, 22nd July 2014; to Peak Rail, Rowsley, 28th July 2014; to Great Central Railway, Loughborough, for gala, 27th August 2014; to Peak Rail, Rowsley, 3rd September 2014; to Andrew Briddon, Darley Dale, 1st May 2015; to Churnet Valley Railway, Cheddleton, 26th May 2016; to Andrew Briddon, Darley Dale, 1st March 2017; to Colne Valley Railway, 25th May 2017; to Old Oak Common Depot, London, for open day held on 2nd September 2017; returned to Colne Valley Railway, Castle Hedingham, 11th September 2017; to Andrew Briddon, Darley Dale, 21st March 2019.

D9525 Swindon 1964 50B 4/68 P D9525

to NCB Philadelphia Locomotive Shed, County Durham; despatched from 50B Dairycoates Depot, Hull, 28th November 1968; to Burradon Colliery, Dudley, 7th March 1975; to Ashington Colliery, 14th March 1975; to Backworth Colliery, 15th December 1975; to Ashington Colliery, 15th August 1980; to Weetslade Coal Preparation Plant, January 1981; to Ashington Colliery, 24th April 1981; to Lambton Engine Works, Philadelphia, 25th July 1983; to Ashington Colliery, 7th February 1984; sold to Kent & East Sussex Railway, Tenterden, 29th September 1987; to Great Central Railway, Ruddington, Nottingham, 26th June 2000; to Barrow Hill Engine Shed Society, Staveley, 13th August 2001; to Battlefield Line, Shackerstone, 2nd March 2002; to South Devon Railway, Buckfastleigh, 27th May 2004; to Peak Rail, Rowsley, 13th April 2005.

D9526 Swindon 1964 86A 11/68 P D9526

to APCM, Westbury, Wiltshire, January 1970; to APCM, Dunstable, 28th May 1971; to APCM, Westbury, 24th November 1971; to West Somerset Railway, Minehead, 3rd April 1980; to East Lancashire Railway, Bury, for gala, 14th July 2014; to West Somerset Railway, Minehead, September 2014; to South Devon Railway, Buckfastleigh, for gala, 7th October 2018; to West Somerset Railway, Minehead, 6th November 2018.

D9527 Swindon 1965 86A 5/69 F No.6 / 9312-94

to NCB Ashington Colliery; despatched from Canton Depot, Cardiff, 30th June 1969; arrived at Ashington, early July 1969; to Lambton Engine Works, Philadelphia, 28th May 1977; to Ashington Colliery, 20th September 1977; to Lambton Engine Works, Philadelphia, 5th September 1978; to Ashington Colliery, October 1978; withdrawn November 1983; scrapped on site by D. Short Ltd of North Shields, January 1984.

D9528 Swindon 1965 86A 3/69 F No.2 / 9312-100

to NCB Ashington Colliery, March 1969; despatched from Canton Depot, Cardiff, 6th March 1969; to BR Gosforth Depot, Newcastle upon Tyne, for repairs, 11th March 1976; to Ashington Colliery by May 1976; to Lambton Engine Works, Philadelphia, June 1977; to Ashington Colliery, by November 1977; scrapped, December 1981.

D9529 Swindon 1965 50B 4/68 P D9529

to Stewarts & Lloyds Minerals Ltd, Buckminster Quarries, Lincolnshire, August 1968; seen at BR Grantham Station, 31st August 1972; to Corby Quarries, 6th September 1972; to BSC Steelworks Disposal Site, Corby, 29th December 1980; to North Yorkshire Moors Railway, Grosmont, 16th March 1981; to Great Central Railway, Loughborough, 11th December 1984; to BR Coalville Depot, open day, 5th June 1988; returned to Great Central Railway; to Nene Valley Railway, Wansford, 8th December 1988; to Battlefield Line, Shackerstone, on loan, 10th April 1995; to Nene Valley Railway, Wansford, 9th November 1995; to Kent & East Sussex Railway, Tenterden, Kent, 23rd June 2000; to Channel Tunnel Rail Link, Beechbrook Farm, near Ashford, on hire, 10th July 2001; to Nene Valley Railway, Wansford, for repairs, April 2002; returned to Channel Tunnel Rail Link, on hire, 28th June 2002; to Tilbury Docks, on hire, 2nd January 2003; to Medway Ports, Chatham Dockyard, on hire, 1st March 2003; to Nene Valley Railway, Wansford, 14th July 2003; to Channel Tunnel Rail Link, Dagenham, on hire, 19th July 2004; to Nene Valley Railway, Wansford, 19th September 2004; to EWS Toton Depot, for tyre turning, 17th November 2004; to Nene Valley Railway, Wansford, 29th November 2004; to Channel Tunnel Rail Link, Dagenham, on hire, 23rd February 2005; to Channel Tunnel Rail Link, Swanscombe, on hire, September 2005; returned to Nene Valley Railway, Wansford, 14th April 2006; to Channel Tunnel Rail Link, Dagenham, on hire, 7th September 2006; to Nene Valley Railway, Wansford, for repairs, 29th September 2006; to Channel Tunnel Rail Link, Swanscombe, on hire, 24th January 2007; to Nene Valley Railway, Wansford, for repairs, 24th April 2007; to Kent & East Sussex Railway, Tenterden, 30th May 2007; to Nene Valley Railway, Wansford, 29th December 2008; to Aggregate Industries Ltd, Bardon Hill, on hire, 7th January 2009; to Nene Valley Railway, Wansford, 9th October 2010.

D9530 Swindon 1965 86A 10/68 F NFT

to Gulf Oil Co Ltd, Cardiff Docks site, for short period, about September 1969; to Gulf Oil Co Ltd, Waterston, Pembrokeshire, 26th September 1969; to BR Swindon Works, for overhaul, 5th August 1971; returned to Gulf Oil, 7th October 1971; to NCB Mardy Colliery, Glamorgan, via BR Canton Depot, Cardiff, after 16th November 1975; to BR Canton Depot and BR Ebbw Junction Depot, for repairs, July to 16th August 1976; to Mardy Colliery, August 1976; to BR Canton Depot, Cardiff, for open day held on 1st October 1977; to Mardy Colliery, about December 1977; scrapped, March 1982.

D9531 Swindon 1965 86A 12/67 P D9531 / ERNEST

to W.H. Arnott Young & Co Ltd, Parkgate, Rotherham; despatched from 85A Worcester Depot, 18th July 1968; re-sold to Hargreaves Industrial Services Ltd, NCBOE British Oak Disposal Point, Crigglestone, November 1968; to NCB Burradon Colliery, Dudley, 10th October 1973; to Ashington Colliery, by March 1974; to Lambton Engine Works, Philadelphia, 26th September 1981; to Ashington Colliery, 27th October 1981; sold to C.F. Booth Ltd of Rotherham, summer 1987; did not leave Ashington site and was re-sold by Booth's for preservation; to East Lancashire Railway, Bury, 3rd October 1987; to North Norfolk Railway, Sheringham, for gala, 11th July 2014; returned, July 2014; to Severn Valley Railway, Bridgnorth, for gala, 28th September 2015; to East Lancashire Railway, Bury, 8th October 2015.

D9532 Swindon 1965 50B 4/68 F 57 / 8311-37

to Stewarts & Lloyds Minerals Ltd, Corby Quarries, November 1968; to BSC Steelworks Disposal Site, Corby, 29th December 1980; scrapped on site by Shanks & McEwan Ltd of Corby, February 1982.

D9533 Swindon 1965 50B 4/68 F 47 / 8311-26

to Stewarts & Lloyds Minerals Ltd, Corby Quarries, December 1968; to BSC Steelworks Disposal Site, Corby, 29th December 1980; scrapped on site by Shanks & McEwan Ltd of Corby, September 1982.

D9534 Swindon 1965 50B 4/68 F ECCLES

to APCM, Hope Cement Works, Derbyshire, October 1968; sold for export (see Appendix C); left Hope on 5th May 1975.

D9535 Swindon 1965 86A 12/68 F 37 / 9312-59

to NCB Ashington Colliery, November 1970; to Burradon Colliery, Dudley, January 1971; to Ashington Central Workshops, about July 1973; to Burradon Colliery, Dudley, about March 1974; to Weetslade Coal Preparation Plant, 5th January 1976; to Backworth Colliery, about May 1976; to Ashington Colliery, 13th September 1980; to Lambton Engine Works, Philadelphia, 5th December 1980; to Ashington Colliery, 23rd April 1981; withdrawn November 1983; scrapped on site by D. Short Ltd of North Shields, January 1984.

D9536 Swindon 1965 87E 5/69 F No.5 / 9312-91

to NCB Ashington Colliery; despatched from Canton Depot, Cardiff, 6th March 1970; to BR Gosforth Depot, Newcastle upon Tyne, for tyre turning, August 1973; returned to Ashington Colliery; to Lambton Engine Works, Philadelphia, 28th January 1977; to Ashington Colliery, 15th May 1977; to Lambton Engine Works, Philadelphia, January 1978; to Ashington Colliery, June 1978; to Lambton Engine Works, Philadelphia, 16th September 1981; to Ashington Colliery, January 1982; scrapped on site, week-ending 30th November 1985.

D9537 Swindon 1965 50B 4/68 P D9537

to Stewarts & Lloyds Minerals Ltd, Corby Quarries, November 1968; to BSC Penn Green Crane Depot, for storage, about November 1981; to Gloucestershire Warwickshire Railway Society, Toddington, 23rd November 1982; to John Scholes, The Old Station,

Rippingale, Lincolnshire, for storage, 8th May 2003; to East Lancashire Railway, Bury, 4th March 2013; to Dean Forest Railway, Lydney, 24th August 2015; returned to East Lancashire Railway, Bury, September 2015; to Ribble Steam Railway, Preston, for gala, 30th September 2015; to East Lancashire Railway, Bury, 6th October 2015; to Spa Valley Railway, Tunbridge Wells, 19th July 2016; returned to East Lancashire Railway, Bury; to Ecclesbourne Valley Railway, Wirksworth, for gala, 15th March 2017; returned to East Lancashire Railway, Bury, 22nd March 2017; to North Norfolk Railway, Sheringham, June 2018; to East Lancashire Railway, Bury, 10th September 2018; to Ecclesbourne Valley Railway, Wirksworth, on hire, 26th October 2018; to Great Central Railway, Loughborough, for gala, 10th April 2019; returned to Ecclesbourne Valley Railway, Wirksworth, by 26th April 2019; to Great Central Railway, for gala, 10th April 2019; to Ecclesbourne Valley Railway, Wirksworth, by 25th April 2019.

D9538 Swindon 1965 87E 5/69 F 160

to Shell-Mex & BP Ltd, Shell Haven, Essex, April 1970; to BR Swindon Works, for overhaul, by 5th August 1970; resold to British Steel Corporation; to BSC Ebbw Vale Steelworks, Monmouthshire, 22nd February 1971; to BSC Corby Quarries, by 25th April 1976; scrapped on site by Shanks & McEwan Ltd of Corby, September 1982.

D9539 Swindon 1965 50B 4/68 P D9539

to Stewarts & Lloyds Minerals Ltd, Corby Quarries, October 1968; to BSC Steelworks Disposal Site, Corby, 29th December 1980; to Gloucestershire Warwickshire Railway, Toddington, 23rd February 1983; to Ribble Steam Railway, Preston, 26th July 2005; to EWS Crewe Depot, for tyre turning, 16th February 2009; to Ribble Steam Railway, Preston, 27th April 2009; to East Lancashire Railway, Bury, 24th April 2014; to Ribble Steam Railway, Preston, 7th August 2014; to Spa Valley Railway, Tunbridge Wells, 11th June 2015; to Epping Ongar Railway, Essex, 16th September 2015; to Ribble Steam Railway, Preston, 21st September 2015; to Peak Rail, Rowsley, on loan, 5th May 2016; to Ribble Steam Railway, Preston, 12th August 2016.

D9540 Swindon 1965 50B 4/68 F 36 / 508 / 2233-508

to NCB Philadelphia Locomotive Shed, County Durham; despatched from 50B Dairycoates Depot, Hull, 29th November 1968; to Burradon Colliery, Dudley, 25th November 1971; to Ashington Colliery, June 1972; to Burradon Colliery, Dudley, by April 1974; to Weetslade Coal Preparation Plant, 3rd January 1976; to Ashington Colliery, 24th April 1981; withdrawn November 1983; scrapped on site by D. Short Ltd of North Shields, 11th January 1984.

D9541 Swindon 1965 50B 4/68 F 66 / 8411-26

to Stewarts & Lloyds Minerals Ltd, Harlaxton Quarries, Lincolnshire, November 1968; to Corby Quarries, August 1974; to BSC Steelworks Disposal Site, Corby, 29th December 1980; scrapped on site by Shanks & McEwan Ltd of Corby, August 1982.

D9542 Swindon 1965 50B 4/68 F 48 / 8311-27

to Stewarts & Lloyds Minerals Ltd, Corby Quarries, December 1968; to BSC Steelworks Disposal Site, Corby, 29th December 1980; scrapped on site by Shanks & McEwan Ltd of Corby, August 1982.

D9544 Swindon 1965 50B 4/68 F D9544
to Stewarts & Lloyds Minerals Ltd, Corby Quarries, 2nd November 1968; dismantled and used for spares from 1970; remains scrapped, September 1980.

D9545 Swindon 1965 50B 4/68 F D9545
to NCB Ashington Colliery, November 1968; later dismantled and used for spares; remains scrapped, early July 1979.

D9547 Swindon 1965 50B 4/68 F 49 / 8311-28
to Stewarts & Lloyds Minerals Ltd, Corby Quarries, December 1968; to BSC Steelworks Disposal Site, Corby, 29th December 1980; scrapped on site by Shanks & McEwan Ltd of Corby, August 1982.

D9548 Swindon 1965 50B 4/68 F 67 / 8411-27
to Stewarts & Lloyds Minerals Ltd, Harlaxton Quarries, Lincolnshire, November 1968; to Corby Quarries, August 1974; to BSC Steelworks Disposal Site, Corby, 29th December 1980; to Hunslet Engine Co Ltd, Leeds, 19th November 1981; overhauled and rebuilt to 5ft 6in gauge; exported to Bilbao, Spain, via Goole Docks, June 1982.

D9549 Swindon 1965 50B 4/68 F 64 / 8311-33
to Stewarts & Lloyds Minerals Ltd, Corby Quarries, November 1968; to Glendon East Quarries, October 1973; to Corby Quarries, 26th June 1974; to BSC Tube Works, Corby, September 1980; to Hunslet Engine Co Ltd, Leeds, 14th November 1981; overhauled and rebuilt to 5ft 6in gauge; exported to Bilbao, Spain, via Goole Docks, June 1982.

D9551 Swindon 1965 50B 4/68 P D9551
to Stewarts & Lloyds Minerals Ltd, Corby Quarries, December 1968; to BSC Tube Works, Corby, July 1980; to West Somerset Railway, Minehead, 5th June 1981; to Royal Deeside Railway, Banchory, 9th November 2000; to Severn Valley Railway, Bridgnorth, 25th November 2013; to Didcot Railway Centre, Oxfordshire, 9th March 2019; to Severn Valley Railway, Bridgnorth, 14th May 2019.

D9552 Swindon 1965 50B 4/68 F 59 / 8411-21
to Stewarts & Lloyds Minerals Ltd, Buckminster Quarries, Lincolnshire, September 1968; seen at BR Grantham Station, 31st August 1972; to Corby Quarries, 6th September 1972; scrapped, September 1980.

D9553 Swindon 1965 50B 4/68 P D9553 / 54
to Stewarts & Lloyds Minerals Ltd, Corby Quarries, November 1968; to BSC Steelworks Disposal Site, Corby, 29th December 1980; to Gloucestershire Warwickshire Railway, Toddington, 23rd February 1983; to Moveright International, Wishaw, Warwickshire, 14th December 2015; to Vale of Berkeley Railway, Sharpness, 6th January 2016; to Berkeley Power Station, Gloucestershire, September 2017.

D9554 Swindon 1965 50B 4/68 F 58 / 8311-38
to Stewarts & Lloyds Minerals Ltd, Corby Quarries, November 1968; to Penn Green Works, for repairs, by 23rd March 1976; to BSC Steelworks Disposal Site, Corby, 29th December 1980; scrapped on site by Shanks & McEwan Ltd of Corby, August 1982.

D9555 Swindon 1965 87E 5/69 P D9555

to NCB Burradon Colliery, Dudley; despatched from Canton Depot, Cardiff, 6th March 1970; to Ashington Colliery, 7th February 1975; to Burradon Colliery, Dudley, March 1975; to Ashington Colliery, by 21st November 1975; to Backworth Colliery, 3rd December 1975; to Ashington Colliery, 15th August 1980; to Rutland Railway Museum, Cottesmore, 24th September 1987; to Northampton & Lamport Railway, on loan, August 1998; to Rutland Railway Museum, Cottesmore, 27th October 1998; to Old Oak Common Depot, London, open day, August 2000; returned to Cottesmore; to Dean Forest Railway, Lydney, about March 2002; to EWS Toton Depot, for tyre turning, 10th June 2003; returned to Dean Forest Railway, Lydney, 11th June 2003; to East Lancashire Railway, Bury, for gala, 9th July 2014; to Dean Forest Railway, Lydney, 30th July 2014.

SECTION 20

LMSR and British Railways built 0-6-0 diesel electric locomotives, numbered 12033-12138, and introduced 1945. Fitted with an English Electric 6KT engine developing 350bhp at 680rpm, and driving wheels of 4ft 0½in diameter. Later classified TOPS Class 11.

12049 Derby 1948 1E 10/71 F 12049

to Day & Sons (Brentford) Ltd, Brentford Town Goods Depot, London, October 1972; to BR Old Oak Common Depot, London, for repairs, October 1976; returned to Day & Sons, January 1977; to Mid Hants Railway, Ropley, Hampshire, 28th July 1998; suffered from fire damage, 26th July 2010; to European Metal Recycling, Kingsbury, for scrap, 3rd November 2010; scrapped July 2011.

12050 Derby 1949 9A 7/70 F 12050

to NCB Philadelphia Locomotive Shed, County Durham; despatched from BR Newton Heath Depot, April 1971; used for spares, June 1971; remains scrapped, June 1972.

12052 Derby 1949 5A 6/71 P 12052

to Derek Crouch (Contractors) Ltd, NCBOE Widdrington Disposal Point, Northumberland, 14th December 1971; to Scottish Industrial Railway Centre, Dunaskin, Dalmellington, 2nd October 1988; to SIRC, Minnivey, 8th May 1994; to SIRC Dunaskin, Dalmellington, 11th March 2002; to Caledonian Railway, Brechin, about 8th April 2002.

12054 Derby 1949 6A 7/70 F 12054

to A.R. Adams & Son, Newport; despatched from BR Birkenhead, 15th September 1971; used as a hire locomotive (see Appendix A); scrapped, April 1984.

12060 Derby 1949 9A 2/71 F 512 / 2233-512

despatched from 9D Newton Heath Depot, initially in 17-40 freight working to Healey Mills, 30th March 1971; arrived at NCB Derwenthaugh Locomotive Shed, Blaydon, County Durham, early April 1971; to Philadelphia Locomotive Shed, 16th April 1971; scrapped on site by C.F. Booth Ltd of Rotherham, December 1985.

12061 Derby 1949 8J 10/71 F NPT

to NCB Nantgarw Coking Plant, Treforest; despatched from BR Springs Branch Depot, Wigan (ran 9Z10), 11th December 1972; to BR Canton Depot, Cardiff, for repairs, 4th

December 1974; to Nantgarw Coking Plant, Treforest, December 1974; to BR Swindon Works, for repairs, 10th December 1981; to Nantgarw Coking Plant, Treforest, 8th February 1982; to Vale of Neath Railway Society, Aberdulais, 23rd August 1987; to Gwili Railway, Bronwydd Arms, 13th September 1991; to Peak Rail, Rowsley, 11th June 2004; to European Metal Recycling, Attercliffe, Sheffield, for scrap, 27th March 2013; scrapped, 28th March 2013.

12063 Derby 1949 8F 1/72 F 5

to NCB Nantgarw Coking Plant, Treforest; despatched from BR Springs Branch Depot, Wigan (ran 9Z10), 11th December 1972; to BR Canton Depot, Cardiff, for repairs, January 1977; to Nantgarw Coking Plant, Treforest; scrapped, November 1987.

12071 Derby 1950 8F 10/71 F 6

to NCB Nantgarw Coking Plant, Treforest; despatched from BR Springs Branch Depot, Wigan (ran 9Z10), 11th December 1972; to BR Canton Depot, Cardiff, for repairs, by 23rd December 1974; to Nantgarw Coking Plant, Treforest, December 1974; to Canton Depot, Cardiff, for repairs, about November 1976; to Nantgarw Coking Plant, Treforest, 15th November 1976; to BR Ebbw Junction Depot, Newport, for repairs, by 21st March 1977; to BR Swindon Works, for repairs, 6th July 1977; to Nantgarw Coking Plant, Treforest, 2nd October 1977; to BR Swindon Works, for repairs, September 1980; to Nantgarw Coking Plant, Treforest, February 1981; to National Smokeless Fuels, Aberaman Phurnacite Plant, Abercwmboi, July 1987; to C.F. Booth Ltd, Rotherham, 18th July 1990; to South Yorkshire Railway Preservation Society, Meadowhall, Sheffield, 24th August 1992; used for spares, June 1995; remains scrapped at Coopers (Metals) Ltd, Attercliffe, Sheffield, June 1995.

12074 Derby 1950 6A 1/72 F 12074

to Johnsons (Chopwell) Ltd, NCBOE, Swalwell Disposal Point, Whickham, County Durham; despatched from BR Crewe Depot, June 1972; to South Yorkshire Railway Preservation Society (HNRC), Meadowhall, Sheffield, 21st June 1989; to European Metal Recycling, Kingsbury, about June 2001; scrapped, July 2002.

12077 Derby 1950 8F 10/71 P 12077

to Cashmore Ltd, Great Bridge, Staffordshire, September 1973; to Midland Railway, Butterley, Derbyshire, 16th December 1978.

12082 Derby 1950 6G 10/71 P 12049

to Shellstar (UK) Ltd, Ince Marshes, Ellesmere Port; despatched from BR Chester Depot, 27th March 1973; to Manchester Ship Canal Company, Ellesmere Port, on loan, 16th July 1974; to Shellstar, 25th October 1974; to BR Swindon Works, for repairs, 22nd December 1977; to Shellstar, April 1978; to South Yorkshire Railway Preservation Society (HNRC), Meadowhall, Sheffield, December 1991; to Cobra Railfreight, Wakefield, West Yorkshire, on hire, March 1993; returned to SYRPS, 9th December 1994; to RFS (Engineering) Ltd, Doncaster, for repairs, 28th November 1995; to Cobra Railfreight, Wakefield, on hire, 5th April 1996; to RFS (Engineering) Ltd, Doncaster, for repairs, 15th October 1997; to SYRPS, 18th December 1997; to Barrow Hill Engine Shed Society (HNRC), Staveley, 24th June 1999; Railtrack registered 01553; to Wabtec, Doncaster, 21st December 2000; to LaFarge Cement, Hope Cement Works, Derbyshire, on hire, 6th

August 2003; to Barrow Hill Engine Shed Society, Staveley, by May 2004; to Whitemoor Yard, March, on hire, 29th September 2004; but unsuitable and returned to Barrow Hill Engine Shed Society, Staveley, 30th September 2004; to Midland Railway, Butterley, 2004; to HNRC, Long Marston, July 2005; to Deanside Transit, Glasgow, on hire, January 2008; to Barrow Hill Engine Shed Society (HNRC), Staveley, September 2010; to Mid Hants Railway, Ropley, 1st November 2010; re-numbered 12049 to replace their original 12049 (which see) that suffered fire damage.

12083 Derby 1950 12A 10/71 P 12083 / M413
seen at Carlisle Kingmoor Depot, 28th April 1973 and 9th June 1973; to Tilcon Ltd, Swinden Lime Works, Grassington, July 1973; to BR Doncaster Depot, for repairs, September 1974; returned to Tilcon Ltd, October 1974; to South Yorkshire Railway Preservation Society (HNRC), Meadowhall, Sheffield, 21st May 1998; to Battlefield Line, Shackerstone, 1st August 2001; sold by HNRC to a preservation group at Shackerstone, November 2006.

12084 Derby 1950 5A 5/71 F 514 / 2233-514
to NCB Burradon Colliery, Dudley, Northumberland, October 1971; to Philadelphia Locomotive Shed, County Durham, 25th November 1971; to Silksworth Colliery, Sunderland, 7th April 1972; to Hylton Colliery, Castletown, July 1972; to Philadelphia Locomotive Shed, 3rd March 1975; to Easington Colliery, 22nd December 1975; to Blackhall Colliery, 5th January 1976; to Bates Colliery, Blyth, 5th April 1976; to Lambton Engine Works, Philadelphia, 25th February 1983; to Philadelphia Locomotive Shed, 21st October 1983; used for spares; remains scrapped on site by C.F. Booth Ltd of Rotherham, November 1985.

12085 Derby 1950 12A 5/71 F 12085
seen at Carlisle Kingmoor Depot, 28th April 1973; to Thos. W. Ward Ltd, Barrow-in-Furness, May 1973; scrapped, about June 1976.

12088 Derby 1951 8J 5/71 P 12088
to Johnsons (Chopwell) Ltd, NCBOE Swalwell Disposal Point, Whickham, County Durham; despatched from BR Springs Branch Depot, Wigan, July 1972; to South Yorkshire Railway Preservation Society (HNRC), Meadowhall, Sheffield, June 1989; to Johnsons (Chopwell) Ltd, Widdrington Disposal Point, Northumberland, on hire, 28th May 1996; to Steadsburn Opencast Site / Disposal Point, from 2007 (west of Widdrington and operative from mid-2007 to 2011); to Butterwell Disposal Point, near Linton, Northumberland, 8th November 2011; to Aln Valley Railway, Alnwick, 6th December 2012.

12093 Derby 1951 5A 5/71 P 12093
to Derek Crouch (Contractors) Ltd, NCBOE Widdrington Disposal Point, Northumberland, 15th December 1971; to Scottish Industrial Railway Centre, Dunaskin, Dalmellington, 9th October 1988; to Caledonian Railway, Brechin, about 8th April 2002.

12098 Derby 1952 8F 2/71 F 12098
to NCB Derwenthaugh Locomotive Shed, Blaydon, County Durham; despatched from 9D Newton Heath Depot in 06-30 freight working to Healey Mills, 31st March 1971; to Philadelphia Locomotive Shed, 16th April 1971; to National Smokeless Fuels Ltd, Lambton Coking Plant, September 1985; to Stephenson Railway Museum, Middle Engine

Lane, North Shields, 5th January 1987; to South Yorkshire Railway Preservation Society (HNRC), Meadowhall, Sheffield, 9th December 1997; to European Metal Recycling, Kingsbury, about June 2001; scrapped, June 2006.

12099 Derby 1952 1E 7/71 P 12099
to Murphy Bros Ltd, NCBOE Lion Disposal Point, Blaenavon, April 1972; to Taylor Woodrow Construction Ltd, NCBOE Cwm Bargoed Disposal Point, 23rd October 1975; to Hargreaves Industrial Services Ltd, NCBOE British Oak Disposal Point, Crigglestone, August 1981; to NCBOE Bowers Row Disposal Point, Astley, 11th February 1983; disused by August 1988; to C.F. Booth Ltd, Rotherham, 21st February 1989; to Severn Valley Railway, Bridgnorth, 26th March 1990.

12119 Darlington 1952 50B 11/68 F 509 / 2233-509
to NCB Philadelphia Locomotive Shed, County Durham; despatched from 50B Dairycoates Depot, Hull, 6th February 1969; to Lambton Engine Works, Philadelphia, November 1980; to Philadelphia Locomotive Shed, January 1981; scrapped on site by C.F. Booth Ltd of Rotherham, November 1985.

12120 Darlington 1952 50B 12/68 F 510
to NCB Philadelphia Locomotive Shed, County Durham; despatched from 50B Dairycoates Depot, Hull, 6th February 1969; to Whittle Colliery, Newton-on-the-Moor, June 1978; to Lambton Engine Works, Philadelphia, August 1979; used for spares; remains scrapped on site by L. Marley of Stranley, March 1980.

12122 Darlington 1952 40B 7/71 F 12122
to Murphy Bros Ltd, NCBOE Lion Disposal Point, Blaenavon, 30th January 1972; suffered collision damage, February 1972; to Taylor Woodrow Construction Ltd, NCBOE Cwm Bargoed Disposal Point, October 1975; to Hargreaves Industrial Services Ltd, NCBOE British Oak Disposal Point, Crigglestone, August 1981; used for spares; remains scrapped on site by Rawden of Barnsley, October 1985.

12131 Darlington 1952 30A 3/69 P 12131
to NCB Betteshanger Colliery, Kent, 25th March 1969; seen at Betteshanger numbered 1802/B3, 15th October 1972; to Snowdown Colliery, Kent, 22nd June 1976; to North Norfolk Railway, Sheringham, 25th April 1982.

12133 Darlington 1952 40B 1/69 F 511 / 2100-526
to NCB Philadelphia Locomotive Shed, County Durham; despatched from 40B Immingham Depot, 9th May 1969; to Lambton Engine Works, Philadelphia, 1979; to Whittle Colliery, Newton-on-the-Moor, about March 1981; to Lambton Engine Works, Philadelphia, 23rd April 1981; to Philadelphia Locomotive Shed, 13th August 1981; scrapped on site by C.F. Booth Ltd of Rotherham, November 1985.

SECTION 21

British Railways built 0-6-0 diesel electric locomotives, numbered 15211-15236, and introduced 1949. Fitted with an English Electric 6KT engine developing 350bhp at 680rpm, and driving wheels of 4ft 6in diameter. Later classified TOPS Class 12.

15222 Ashford 1949 73C 10/71 F 15222

to Cashmore Ltd, Newport, June 1972; to John Williams Ltd, Blaenyfan, Kidwelly, 1974; used as a stationary generator; scrapped on site, September 1978.

15224 Ashford 1949 75C 10/71 P 15224

to NCB Betteshanger Colliery, Kent; despatched from BR Brighton, October 1972; seen at Betteshanger numbered 1802/B5, 15th October 1972; to Snowdown Colliery, Kent, 27th May 1976; left Snowdown Colliery, 9th October 1982; stored in BR Hove Goods Yard, October 1982; to Brighton Works Locomotive Association, Preston Park Car Sheds, Brighton, April 1983; to Lavender Line, Isfield, East Sussex, June 1985; to Spa Valley Railway, Tunbridge Wells, 21st January 1998 to 1st February 1998; to Lavender Line, Isfield, 1st February 1998; to Spa Valley Railway, Tunbridge Wells, 2011.

15231 Ashford 1951 73F 10/71 F TILCON

to Tilcon Ltd, Swinden Lime Works, Grassington, June 1972; scrapped on site, February 1984.

SECTION 22

Ruston & Hornsby Ltd built, 3ft 0in gauge, 4-wheel diesel mechanical locomotive, number ED10, built in 1958 (Ruston's class 48DS). Fitted with a Ruston 4YC engine developing 48bhp at 1375rpm, three speed gearbox, and driving wheels of 2ft 6in diameter. No TOPS classification.

ED10 RH 411322 1958 BSD 2/65 P E9

used from new in Departmental service by British Railways to push narrow gauge bolster wagons loaded with wooden sleepers into the creosote impregnation chambers, at Beeston Sleeper Works; system closed, early 1965; to Thos. W. Ward Ltd, Sheffield, February 1965; to Cleveland Bridge & Engineering Co Ltd, Darlington, May 1966; used by CB&E on the Tinsley Viaduct, Sheffield, contract; to Shephard Hill & Co Ltd (contractors), February 1970; fitted with rubber tyres and stabilisers and used on a contract to construct three miles of hover-train track for Tracked Hovercraft Ltd, Earith, Cambridgeshire, from 1971; project cancelled in 1973 and site closed 7th September 1974; to E. Hampton, Church Farm, Fenstanton, St Ives, Huntingdonshire, for preservation, 1975; to Irchester Narrow Gauge Railway Trust, Irchester Goods Shed, Northamptonshire, 28th September 1987; to Irchester Country Park, Northamptonshire, 8th June 1988; re-gauged to metre gauge, 5th May 1991.

SECTION 23

Ruston & Hornsby Ltd built, 1ft 6in gauge, 4-wheel diesel mechanical locomotive, number ZM32, built 1957 (Ruston's class LAT). Fitted with a Ruston 2VSH engine developing 20bhp at 1200rpm, two speed gearbox, and driving wheels of 1ft 4¼in diameter. No TOPS classification.

ZM32 RH 416214 1957 ZJ 3/64 P ZM32 / HORWICH / 11

to S.E.E.C. Manchester, September 1965; sold to a buyer in British Honduras, but sale cancelled and locomotive stored at Liverpool Docks; resold for preservation; to R.P. Morris, Longfield, Kent, December 1971; to Alan Keef Ltd, Cote, Oxfordshire, 17th April 1973; rebuilt to 2ft 0in gauge; to Narrow Gauge Railway Centre of North Wales, Gloddfa Ganol, Blaenau Ffestiniog, Gwynedd, 20th July 1976; to R.P. Morris, Blaenau Ffestiniog, 1998; to FMB Engineering, Oakhanger, Hampshire (for repairs and conversion back to 1ft 6in gauge), 6th October 1999; to Uppertown, near Ashover, Derbyshire, for storage, 30th May 2000; to Steeple Grange Light Railway, Wirksworth, Derbyshire, 9th June 2000; to Whaley Bridge, for overhaul, 24th November 2002; to Steeple Grange Light Railway, Wirksworth, Derbyshire, 15th January 2003.

SECTION 24

English Electric Ltd built, 0-6-0 diesel electric locomotives, built 1956. Fitted with an English Electric 6RKT engine developing 500bhp at 750rpm, and driving wheels of 4ft 0in diameter. D0226 was diesel-electric transmission, and D0227 was diesel-hydraulic. These locomotives were tested by British Railways but were never incorporated into capital stock. No TOPS classification.

D0226	**EE**	**2345**	**1956**	**- -**	**12/60**	**P**	**D0226 / VULCAN**
	VF	**D226**					

Given trials on British Railways from 1956 to December 1960; mainly based at Stratford Depot in London; originally numbered D226 to match its works number but re-numbered D0226 in August 1959 to avoid clash of number with new Type 4 (later Class 40) locomotive; withdrawn, 31st December 1960; returned to English Electric Ltd, Vulcan Works, Newton-le-Willows, January 1961; stored; to Keighley & Worth Valley Railway, Haworth, March 1966; to BR Doncaster Depot, for tyre turning, 4th March 1979; returned to KWVR, Haworth, 1979; to Railfest, York, for display, 26th May 2004; returned to KWVR, Haworth, June 2004.

D0227	**EE**	**2346**	**1956**	**- -**	**9/59**	**F**	**D0227 / BLACK PIG**
	VF	**D227**					

Given trials on British Railways from 1956 to 1960; mainly based at Stratford Depot in London; originally numbered D227 to match its works number but re-numbered D0227 in August 1959 to avoid clash of number with new Type 4 (later Class 40) locomotive; returned to English Electric Ltd, 1960; to Robert Stephenson & Hawthorns Ltd, Darlington 1960; seen on 13th August 1964; scrapped soon after.

SECTION 25

Ruston & Hornsby Ltd built, 4-wheel diesel mechanical locomotives (Ruston's class LB). Fitted with Ruston 3VSH engines developing 31bhp at 1800rpm, and 1ft 4¼in diameter wheels. Ruston & Hornsby built no less than 557 examples of this class, to 23 different narrow gauges, with the two examples recorded below being of 2ft 0in gauge. No TOPS classification.

85049 RH 393325 1956 CJ c4/86 P 85049

used from new by British Railways at the extensive Chesterton Junction Permanent Way Materials Depot, Cambridge, which was set up in the 1950s; the 2ft gauge system was used to convey reclaimed track fittings to various parts of the yard in flat wagons and skips; to Northamptonshire Ironstone Railway Trust, Hunsbury Hill, 2nd August 1986; to Overland Railways, Chidham, near Chichester, for restoration, 1989; to Vobster Light Railway, Holwell Farm, Mells, Somerset, 13th January 1992; to Somerset & Avon Railway Company, Radstock, 25th June 1994; to Derbyshire Dales Narrow Gauge Railway, Rowsley, Derbyshire, February 1999; to Nunckley Narrow Gauge Railway, Rothley, Leicestershire, 15th March 2017.

85051 RH 404967 1957 CJ c4/86 P NPT

used from new by British Railways at Chesterton Junction Permanent Way Materials Depot, Cambridge (as detailed above); to Cadeby Rectory, Market Bosworth, Leicestershire, 3rd July 1986; to Derbyshire Dales Narrow Gauge Railway, Rowsley, Derbyshire, 6th May 2006.

SECTION 26

Ruston & Hornsby Ltd built, 4-wheel diesel mechanical locomotive (Ruston's class 48DL). Fitted with a Ruston 4VRO engine developing 48bhp, and 1ft 6in diameter wheels. Ruston & Hornsby built 1,127 of this class, to forty different gauges, with works number 221615 (see Appendix C) being of 2ft 0in gauge and 224337 of 3ft 0in gauge. Neither locomotive was ever allocated a BR number. No TOPS classification.

- RH 224337 1945 LSD 1964 P 06/22/6/2

used from new (named MONTY) by British Railways to serve the creosoting plant and workshops at Lowestoft Sleeper Depot, Suffolk; works closed, May 1964; to dealer A. King & Sons Ltd, Norwich, September 1964; to Lynite Concrete Co Ltd, Bury Road, Ramsey, date not known; used to transport concrete products from autoclaves to storage sheds; use of railway ceased in 1974; stored for fourteen years; to J. & K. Harris, scrapyard, Norwood Road Industrial Estate, March, Cambridgeshire, December 1988; purchased by Andrew Wilson, 1995; to Green's Industrial Services, Sibthorpe, Nottinghamshire, for storage, 14th June 1995; to Andrew Wilson, Leeds, 27th March 1997.

SECTION 27

Ruston & Hornsby Ltd built 0-6-0 diesel electric locomotives (Ruston's class 165DE). British Railways purchased five for Departmental work on the Western Region which were initially numbered PWM650 to PWM654. Fitted with a Ruston 6VPH engine developing 155hp, and driving wheels of 3ft 2½in diameter. Later re-numbered 97650 to 97654. No TOPS classification.

PWM650 RH 312990 1952 81D 4/87 P PWM650 / 97650

97650 to Lincoln City Council, Holmes Yard, Lincoln, 27th February 1991; to Appleby Frodingham Railway Preservation Society, Scunthorpe, for storage, 4th February 1994;

to Lincolnshire Wolds Railway, Ludborough, by March 1995; to Peak Rail, Rowsley, 11th January 2017.

PWM651 RH 431758 1959 86A 9/98 P PWM651
97651 to Northampton & Lamport Railway, Chapel Brampton, 10th November 1998; to Strathspey Railway, Aviemore, May 2008; to Swindon & Cricklade Railway, 14th August 2015.

PWM653 RH 431760 1959 81D ? F 97653
97653 to Yorkshire Engine Company, Long Marston; despatched from Reading Depot, 6th November 1998; allocated YEC works number L163; acquired by John Payne from receivers of YEC, November 2001; locomotive remained at Long Marston; used for spares; remains to Brian Hirst Recycling, Bullington Cross, Hampshire, 1st August 2011; scrapped, 2011.

PWM654 RH 431761 1959 ? ? P PWM654
97654 latterly used at On-Track Machine Depot, Slateford, Edinburgh; to Peak Rail, Rowsley; despatched from Slateford, 6th April 2005.

SECTION 28

Ruston & Hornsby Ltd built, 4-wheel diesel mechanical locomotive (Ruston's class 20DL). Fitted with a Ruston 2VSO engine developing 20hp, and wheels of 1ft 4½in diameter. Ruston & Hornsby built 1,198 of this class, to 37 different gauges, with works number 202005 being of 2ft 3in gauge. This locomotive was never allocated a British Railways number. No TOPS classification.

—- RH 202005 1940 HHC ? F ?
used by British Railways on an internal system around the timber stockpiles at its Sleeper Depot at Hall Hills, close to Boston Docks; locomotive acquired new in 1940; to John S. Allen & Son Ltd, Mardyke Works, Cranham, near Upminster, Essex (via Rundle & John Philips & Co Ltd), by May 1967; exported to Singapore, date not known; subsequent history not known; believed scrapped.

LOCOMOTIVE APPENDICES

APPENDIX A : A.R. ADAMS & SON, NEWPORT

According to a Newport trade directory advertisement of the 1930s the Adams business was established in 1893. The directory describes the firm as 'Engineers, General Smiths and Boiler Makers' and its Pill Bank Ironworks stood at the junction of the west side of Robert Street and Courtybella Terrace, Newport (ST312868). The site was served by a branch of the Tredegar Estates Line running along Courtybella Terrace. The rail-connected locomotive repair shop and 'Adams Yard' were opposite the ironworks (ST 313867) between the east side of Robert Street and Price Street. Adams had acquired this site and its private sidings by 1925, and the previously mentioned directory notes that repairs to marine, locomotive and stationary boilers and fireboxes were major activities at that time. Pill Bank Ironworks was relocated in the early 1960s (by April 1962) to Coomassie Street (ST 315864) near the western end of the famous transporter bridge. Thereafter locomotives were stabled and repaired at the United Wagon Company (Barker & Lovering) premises nearby, served by the Tredegar Estates Line, on King's Parade (ST319869). This arrangement ceased about 1972, possibly earlier, and thereafter other premises were used in the area until 1977, believed to be at the end of Portland Street (ST321868). Almost certainly overhauls and sales of locomotives had been a feature of the firm's trade from its early days, although the earliest authenticated locomotive transaction dates from 1920. The practice of hiring, or loaning, a locomotive to a customer (whilst his own engine was overhauled by Adams) seems to have been established early on and was possibly the origin of what became an important locomotive hire trade. Adams' hire business has been fairly well documented by enthusiasts from the 1950s onwards. The company hired out steam locomotives of standard and narrow gauges, whilst the final phase from about 1968 onwards saw them acquire a small fleet of second-hand 0-6-0 diesel locomotives, mostly surplus ex-BR class 03 and 04 machines. In this 'diesel era' the company hired locomotives to various concerns, which were mainly involved with the coal industry in South Wales. Most ex-BR locomotive enthusiasts will now be aware of the Adams company, but unfortunately very few such enthusiasts are known to have actually visited Adams' premises and thus definitive dated sightings are extremely rare. Certain repairs are known to have been contracted out to BR depots at Ebbw Junction (Newport) and Canton (Cardiff). Between hirings, diesel locomotives were stored (possibly from about late 1972) inside what has been described as 'wired-off compounds' adjacent to the workshops and in Dock Street Goods Yard (ST318871). One report describes a compound as: "an 8ft high corrugated iron fence surmounted by barbed wire and quite impregnable!" Certain published reports have referred to a compound as being at Rowecord Engineering Ltd, but it is now understood that one of the compounds was merely adjacent to Rowecord – and that this firm had no connection to Adams. Details of known hirings are given below. It will be noticed that there are many gaps between known hirings, and these periods may be explained by locomotives being stored at Adams' premises in Newport. This is not definite, however, as such gaps may involve hirings which have not yet come to light. A thorn in the side of enthusiasts was Adams' practice of re-painting its diesel locomotives in green or blue livery and not re-applying their original running numbers. D2193 is known to have carried a cab-side plaque reading: "A.R. Adams & Son, Newport, Mon, Locomotive Hire Service". D2186, D2244, D2276 and 12054 are known to have retained their BR numbers in at least the early days of Adams ownership, but blank cab-sides on re-painted locomotives seem to have led to enthusiasts assuming identities (which are not necessarily correct) and in the course of time assumptions took on the status of facts. Certain anomalies have been corrected since 7BRD, but this appendix is a work in progress and any enthusiasts who visited Adams' premises or saw locomotives out on hire are asked to submit dated sightings and photographs to the author.

D2139 sold to Adams, Newport, and delivered direct to NCB Marketing Department, Gwent Coal Concentration Depot, Newport, 10th December 1968; to United Wagon Company, Newport by March 1969, and seen there (painted light green) receiving attention, 23rd March 1969; to Adams, Newport (assumed), about April 1969; to NCB Coal Products Division, Nantgarw Coking Plant, Treforest, July 1969; to Adams, Newport, January 1970; an unidentified 03 (which may have been D2139) was seen at Coed Cae, 29th June 1970; to Nantgarw Coking Plant, Treforest, August 1970 to March 1971; seen at Nantgarw in green livery on 21st August 1970; sold to NCB Coed Ely Coking Plant, Tonyrefail, (received number 1) by 30th March 1971; seen at Coed Ely on 23rd May 1971, 20th August 1971, 5th June 1972, 16th July 1972, 5th November 1972, and 18th June 1974. (now preserved – see main listing.)

D2178 sold to Adams, Newport, and delivered direct to Aberaman Colliery Washery, January 1970; to Wiggins Teape Ltd, Ely Paper Works, Cardiff, 24th February 1970; seen at Wiggins Teape on 28th February 1970, 20th August 1970 (when in green livery), 28th April 1971 (when noted to bear 84A shed plate), plus on 23rd August 1971; to Powell Duffryn Fuels Ltd, NCBOE Gwaun-cae-Gurwen Disposal Point, from about March 1972 to July 1972; seen at Gwaun-cae-Gurwen, 18th July 1972; to Adams, Newport, and seen there on 10th March 1973 and 20th May 1973; sold to Coed Ely Coking Plant, Tonyrefail, May 1974; seen at Coed Ely on 18th June 1974. (now preserved – see main listing.)

D2181 sold to Adams, Newport, and delivered direct to NCB Marketing Department, Gwent Coal Concentration Depot, Newport, 10th December 1968; seen at Gwent CCD, 12th July 1969; sold to Gwent CCD by 23rd January 1970; seen at Gwent CCD on 17th August 1970 (by now in red livery), 12th March 1972, 15th July 1972, 21st December 1972 (by now with PRIDE OF GWENT nameplates), 19th May 1973, 21st December 1973, 11th April 1974, 22nd April 1975, 11th April 1985, 11th June 1985 and 25th April 1986. (later scrapped – see main listing.)

D2182 sold to Adams, Newport, and delivered direct to NCB Coal Products Division, Caerphilly Tar Works, 29th November 1968; worked at Caerphilly to February 1969; to Glyn Neath Disposal Point, on hire (or possibly on trial) when seen on 15th March 1969; shortly after sold to Sir Lindsay Parkinson & Co Ltd and used at Glyn Neath Disposal Point; seen on 26th March 1969, 6th April 1969, 26th May 1969, 3rd June 1969, 5th October 1969, 10th January 1970, 29th August 1970, 31st March 1971 and 29th May 1972; re-purchased by Adams, about July 1972 and moved to NCB Coal Products Division, Caerphilly Tar Works, and seen there on 26th November 1972; believed to have moved to Adams, Newport, by 10th March 1973 (but needs confirmation); to Lindley Plant, Gatewen Disposal Point, New Broughton, September 1973, to whom the locomotive was sold. (now preserved – see main listing.)

D2186 sold to Adams and delivered direct to Aberaman Colliery Washery, 8th February 1970; stayed at Aberaman to about September 1970; seen at Aberaman Colliery Washery in BR livery on 21st August 1970; to Adams, Newport, about September 1970; to Aberaman Colliery Washery, about December 1970; to NCB Tower Colliery, Hirwaun, from March 1971 to about September 1972; seen at Tower, 31st March 1971, 1st July 1971 and 19th August 1971; returned to Adams, Newport, about September 1972; overhauled and painted in green livery; locomotive believed to be D2186 seen at Adams, Newport, on 10th March 1973, 19th May 1973, 20th April 1975, 7th October 1978, 14th October 1979, and 1st June 1980; scrapped, January 1981.

D2193 sold to Adams, Newport, and delivered direct to Powell Duffryn Fuels Ltd, NCBOE Coed Bach Disposal Point, Kidwelly, September 1969; seen at Coed Bach on 10th January 1970; to Adams, Newport, March to June 1970; to NCB Coal Products Division, Coed Ely Coking Plant, Tonyrefail, June 1970; seen at Coed Ely in green livery on 24th August 1970, 3rd November 1970 and 9th December 1970; to NCB Mountain Ash Colliery, from February 1971 to about April 1971; seen at Adams, Newport, 18th August 1971; to Powell Duffryn Fuels Ltd, NCBOE Coed Bach Disposal Point, Kidwelly, early 1972; to NCB Coal Products Division, Nantgarw Coking Plant, Treforest, by 31st May 1972; seen at Monsanto Chemicals Ltd, Newport, 15th July 1972; to Adams, Newport; seen at Adams, Newport, on 10th March 1973 and 20th May 1973; to NCB Taff Merthyr Colliery, Treharris, from August 1973 to 7th December 1973; a locomotive which MAY have been D2193 was seen at NCBOE Coed Bach Disposal Point, Kidwelly, 12th July 1974; to Adams, Newport, April 1975; seen in transit at BR Alexandra Dock Junction sidings, Newport, 20th April 1975; to NCB Garw Colliery, Blaengarw, from October 1977 to 27th September 1978; seen at Garw Colliery on 20th April 1978; returned to Adams, Newport; seen at Adams, Newport, 7th October 1978, 14th October 1979, and 1st June 1980; scrapped, January 1981.

D2244 sold to Adams, Newport, and delivered direct to Monsanto Chemicals Ltd, Newport, August 1970; seen at Monsanto in BR livery on 17th August 1970; stayed at Monsanto to March 1971; to NCB Coed Cae Colliery, Heol-y-Cyw, near Pencoed, by 28th March 1971 to about March 1972; seen at Coed Cae on 30th March 1971, 11th July 1971, 19th August 1971 and 2nd September 1971; to NCB Wern Tarw (seen there 11th September 1971); to NCB Ogmore Central Washery, Ogmore Vale, 16th March 1972; stayed at Ogmore to 19th May 1972; to NCB Gwent Coal Concentration Depot, Newport; seen at Gwent CCD on 15th July 1972 and stayed to December 1972; seen in Newport on 21st December 1972 and 2nd January 1973 (in blue livery and with windows boarded-up), and this believed to be in the Dock Street area; seen at Adams, Newport, 21st December 1973, 22nd August 1974, 20th April 1975, 12th October 1975; scrapped, January 1981.

D2276 sold to Adams, Newport, July 1970; used for spares; seen in a dismantled state at Adams' premises at Newport (in BR livery) on 18th August 1971; 19th March 1972, 15th July 1972, 10th March 1973, 19th May 1973, 10th April 1974, 16th April 1974, 20th April 1975, and 24th September 1976; scrapped, May 1977.

12054 sold to Adams, Newport, and delivered direct to NCB Mountain Ash Colliery, 15th September 1971; seen at Mountain Ash, 18th September 1971; stayed at Mountain Ash

Colliery to May 1972; to NCB Tower Colliery, Hirwaun, from May 1972 to September 1973; seen at Tower Colliery on 16th July 1972 and 5th November 1972; to BR Canton Depot, Cardiff, for repairs, by 1st October 1973; seen at Canton on 10th October 1973 and 3rd November 1973; to Adams, Newport (where repainted in light green livery), early April 1974; despatched from Adams to NCB Mardy Colliery, 9th April 1974; seen at Mardy Colliery on 9th April 1974, 20th April 1975 and 13th May 1975; left Mardy Colliery, 6th October 1975; seen at BR Ebbw Junction Depot, Newport, undergoing repairs, 11th October 1975; to NCB Mardy Colliery, March 1976; seen at Mardy Colliery on 26th March 1976 and 20th April 1978 and stayed to September 1979; to BR Canton Depot, Cardiff, for repairs, September 1979; seen at Canton Depot, 14th October 1979 and 17th November 1979; to NCB Mardy Colliery, from late November 1979 to March 1981; to Adams, Newport, by road, March 1981; seen at Adams' Pillbank Works, Coomassie Street, Newport, 12th August 1981, 11th May 1982, 13th November 1982 and 11th February 1984; scrapped, April 1984.

Adams also owned three industrial 0-6-0 diesels: HE 5673 of 1963, ROGIE (DC 2218/VF D47 of 1947), and GWENT (DC 2252/VF D78 of 1948). The company previously had a large number of industrial steam locomotives of standard and narrow gauges.

The text above shows the ex-BR career of 12054, including a mention that it was seen at Mardy Colliery on 13th May 1975. On that date, 12054 was photographed in the colliery yard with a train of coal wagons. (Photograph by Adrian Booth)

APPENDIX B : T.J. THOMSON & SON LTD, STOCKTON-ON-TEES

Thomson's business appears to have been established as metal merchants in 1871 with an office in Middlesbrough. Edgar Gilkes joined Thomas J. Thomson & Co in 1881 and Thomson & Gilkes then operated from the Millfield Iron Works at a site or sites in Stockton-on-Tees. Gilkes retired in February 1883. In 1928 the business, now titled T.J. Thomson & Son Ltd, was purchased by H.E.I Turner although the company's name remained the same and, in 1932, it moved to occupy the site of the former Moor Steel and Iron Works which it renamed the Millfield Works. These premises were situated on the west side of the LNER line from Yarm to Norton, a short distance south of Stockton Station. After 1940, traffic was exchanged with the railway company at Phoenix Sidings situated between the main line and the scrapyard. Here a Thomson locomotive collected condemned wagons and locomotives and propelled them into the yard where several sidings spread across the site, including some passing under gantries. Over the years Thomson's scrapped numerous locomotives, including a number of Class 31s and, from the 1970s, many industrial diesels. The disappearance of much heavy industry in the area led to the rundown of the scrapyard from 2014. Most of the internal wagons were scrapped and the remaining locomotives sold to Ed Murray & Sons Ltd in 2015. The company continued to trade in scrap but stopped buying materials for processing on site in December 2016 and the remaining track was auctioned on 1st March 2017. Back in May 1970, Thomson's purchased three ex-BR Departmental locomotives, which were despatched to its Millfield Works from BR Thornaby Depot, having been stored at the latter location from about August 1969. They were never used at Thomson's works, but were stored for over eleven years until scrapped in October 1981. The trio (which were all 4-wheel locomotives of the maker's class 88DS) were fitted with 88hp engines and diesel-mechanical transmission.

Departmental number	*Builder*	*Works number*	*Year built*	*Last working location*
56	RH	338424	1955	Etherley Tip, near Bishop Auckland
82	RH	425485	1959	Dinsdale Welded Rail Depot, near Darlington
87	RH	463152	1961	Geneva Yard, Darlington

APPENDIX C : LOCOMOTIVES SOLD ABROAD

The very first confirmed ex-BR locomotives to be exported (but see the mystery locomotive and postscript below) were D3639 and D3649 which were shipped to West Africa in March 1970. Thereafter at least a further thirty-two followed up to June 1982. A major player in this trade was the firm of Shipbreaking (Queenborough) Ltd which had yards at Cairnryan Port (Scotland) and in Kent. During this approximately twelve-year period examples of classes 03, 04, 06, 07, 08, 10 and 14 moved abroad, whilst two narrow gauge Ruston & Hornsby locomotives also fall within the remit of this appendix. Once these thirty-four locomotives were abroad it was extremely rare for any to be seen by British enthusiasts, and reports giving positive sightings have been few and far between. Several locomotives have *never* been reported during the author's forty-five years as compiler of the BRD records. Whilst preparing this volume for publication, a concerted effort has been made to update each entry with all known information. This is often sparse and there are many gaps in our knowledge. Anyone who can provide additional information is asked to contact the author.

D2010 Swindon 1958 51L 11/74 F ?

03010 despatched to Stranraer from Tyne Yard, 18th May 1976; moved by road to Shipbreaking (Queenborough) Ltd, Cairnryan Port works; exported to Italy, May 1976; subsequent history not known; believed scrapped.

D2019 Swindon 1958 32A 7/71 F 1

to Shipbreaking (Queenborough) Ltd, Kent, about June 1972; exported from Sheerness Docks to Stabilimento ISA, Ospitaletto, Brescia, Italy, September 1972; company later known as Acciaierie ISA; still in use, May 1997; disused by 30th August 1999; still there 8th March 2000; subsequent history not known; believed scrapped.

D2032 Swindon 1958 32A 7/71 F 2

to Shipbreaking (Queenborough) Ltd, Kent, about June 1972; exported from Sheerness Docks to Stabilimento ISA, Ospitaletto, Brescia, Italy, August 1972; company later known as Acciaierie ISA; still in use, September 1991; disused by 30th August 1996; still there 1st November 2002; subsequent history not known; believed scrapped.

D2033 Swindon 1958 32A 12/71 F PROFILATINAVE 2

to Shipbreaking (Queenborough) Ltd, Kent, August 1972; exported from Sheerness Docks to Siderurgica SPA, Montirone, Brescia Italy, August 1972; still in use, 6th July 2004; subsequent history not known; believed scrapped.Amendment Lists 18EL

D2036 Swindon 1958 32A 12/71 F PROFILATINAVE 1

to Shipbreaking (Queenborough) Ltd, Kent, about June 1972; exported from Sheerness Docks to Siderurgica SPA, Montirone, Brescia, August 1972; disused by 7th June 1998; dismantled and being used for spares, 6th August 2003; subsequent history not known; believed scrapped.

D2081 Doncaster 1960 31B 12/80 P 03081

03081 despatched by road from BR Swindon Works, 13th November 1981; exported to Sobemai, Maldegem, near Bruges, Belgium, November 1981; to Genappe sugar factory, Belgium, by 20th July 1991; extant but disused, 13th September 1999; returned to England (see main listing).

D2098 Doncaster 1960 51A 11/75 F ?

03098 despatched to Stranraer from Tyne Yard, 18th May 1976; moved by road to Shipbreaking (Queenborough) Ltd, Cairnryan Port yard; exported to Italy, May 1976; a locomotive thought to be this seen at Acciaierie Rumi, Montello, Italy in December 1990; subsequent history not known; believed scrapped.

D2128 Swindon 1960 82A 7/76 P D2128

03128 to Bird's Commercial Motors Ltd, Long Marston, October 1976; resold for export and shipped from Harwich, December 1976; to Zeebouw-Zeezand (where re-engined and numbered 6G2) for Zeebrugge Port Authority harbour extension contract; at Gent Depot (Belgium Railways) for tyre turning and fitting with Deutz V12 engine, March 1989; to Stoomcentrum, Maldegem, May 1989; returned to England (see main listing).

D2134 Swindon 1960 82A 7/76 P 03134

03134 to Bird's Commercial Motors Ltd, Long Marston, October 1976; resold for export, and shipped about January 1977; to Zeebouw-Zeezand (where re-engined and numbered 6G1) for Zeebrugge Port Authority harbour extension contract; at Gent Depot (Belgium Railways) for tyre turning and fitting with Deutz V12 engine, March 1989; to Stoomcentrum, Maldegem, May 1989; returned to England (see main listing).

D2153 Swindon 1960 51L 11/75 F ?

03153 despatched to Stranraer from Tyne Yard, 18th May 1976; moved by road to Shipbreaking (Queenborough) Ltd, Cairnryan Port yard; exported to Italy, May 1976; subsequent history not known; believed scrapped.

D2156 Swindon 1960 52A 11/75 P ?

03156 despatched to Stranraer from Tyne Yard, 18th May 1976; moved by road to Shipbreaking (Queenborough) Ltd, Cairnryan Port yard; exported to Italy, May 1976; seen at Ferramanta Pugliese, Terlizzi, Bari, (in disused condition), 17th May 2006; extant at Terlizzi, 6th February 2019.

D2157 Swindon 1960 50C 12/75 F ?

03157 to Shipbreaking (Queenborough) Ltd, Kent, July 1976; exported from Sheerness Docks to Trieste, Italy, February 1977; initially stored at, or adjacent to, the premises of Acciaieria Ferriera Adriatica, Trieste Docks; later rebuilt by IPE, Verona; to Acciaieria ISA, Chiari steelworks, Brescia; seen at Chiari, September 1996; scrapped on site, 23rd May 1997.

D2164 Swindon 1960 30A 1/76 F ?

03164 to Shipbreaking (Queenborough) Ltd, Kent; despatched on 5th July 1976; arrived on 16th July 1976); exported from Sheerness Docks to Trieste, Italy, February 1977; initially stored at, or adjacent to, the premises of Acciaieria Ferriera Adriatica, Trieste Docks; later rebuilt at IPE, Verona; to Chiari steelworks, Brescia; seen at Chiari, 10th May 1997; scrapped on site, 23rd May 1997.

D2216	**DC**	**2539**	**1955**	**30A**	**5/71**	**F**	**3**
	VF	**D265**					

to Shipbreaking (Queenborough) Ltd, Kent, 1972; seen at Shipbreaking, 11th July 1972; exported from Sheerness Docks to Stabilimento ISA, Ospitaletto, Brescia, Italy, September 1972; company later known as Acciaierie ISA; still in use, April 1992; disused (no engine), 30th August 1996; still present 8th March 2000; subsequent history not known; believed scrapped.

D2231	**DC**	**2555**	**1956**	**16C**	**3/68**	**F**	**?**
	VF	**D281**					

sold to R. E Trem & Company, Finningley, Doncaster, 1969; seen at Etches Park, Derby, 27th March 1970; to Steelbreaking & Dismantling Co, Chesterfield, for storage, about April 1970; noted at Chesterfield, 14th June 1970; exported (possibly via Shipbreaking (Queenborough) Ltd, Kent); to Attilio Rossi, Rome, Italy, about 1971; used on contract for relaying Milan to Como line, 1973; observed in Venice, 11th September 1981; to I.P.E. Locomotori, Verona, 1983; observed in Verona, 31st May 1986; subsequent history not known; believed scrapped.

D2242	**DC**	**2572**	**1956**	**55H**	**10/69**	**F**	**?**
	RSH	**7858**					

sold to R.E. Trem & Co of Finningley, Doncaster; consigned to C.F. Booth Ltd of Rotherham, for storage, by 13th June 1970; still at Booth's on 21st August 1971; to Shipbreaking (Queenborough) Ltd, Kent, unknown date thereafter; exported from Sheerness Docks to Italy, May 1972; seen at Feralpi, Lonato, 1976; subsequent history not known; believed scrapped.

D2289	**DC**	**2669**	**1960**	**70D**	**9/71**	**P**	**NPT**
	RSHD	**8122**					

to Shipbreaking (Queenborough) Ltd, Kent, about April 1972; exported from Sheerness Docks to Acciaierie di Lonato, Brescia, Italy, April 1972; site taken over by Lonato SpA, 2006; left Lonato, 9th June 2018; arrived at Western Docks, Dover, 12th June 2018; to Heritage Shunters Trust, Peak Rail, Rowsley (see main listing), 13th June 2018.

D2295	**DC**	**2675**	**1960**	**70D**	**4/71**	**F**	**?**
	RSHD	**8128**					

to Shipbreaking (Queenborough) Ltd, Kent, 18th March 1972; exported from Sheerness Docks about May 1972; destination not confirmed but may have been Siderurgica Meridionale Stefana Antonio S.P.A. Termoli, in Southern Italy; subsequent history not known; believed scrapped.

D2432	**AB**	**459**	**1960**	**65A**	**12/68**	**F**	**?**

sold to R.E. Trem Ltd of Finningley, Doncaster; re-sold to Shipbreaking (Queenborough) Ltd, Kent; arrived about 20th May 1969; exported from Sheerness Docks to Trieste, Italy, March 1977; initially stored at, or adjacent to, the premises of Acciaieria Ferriera Adriatica, Trieste Docks; subsequent history not known; believed scrapped.

D2993 RH 480694 1962 70D 10/76 F ?
07009 to Shipbreaking (Queenborough) Ltd, Kent, about March 1977; exported from Sheerness Docks to Trieste, Italy, March 1977; initially stored at, or adjacent to, the premises of Acciaieria Ferriera Adriatica, Trieste Docks; to Attilio Rossi, Rome, date not known; subsequent history not known; believed scrapped.

D3047 Derby 1954 70D 7/73 P 105
to BR Derby Works, January 1974; overhauled and modified; exported from Canada Dock, Liverpool, about 17th February 1975; to Lamco Mining Co, Tokadeh Mine, Nimba, Liberia, February 1975; last used 29th November 1981; extant, but observed derelict, December 2011; subsequent history not known.

D3092 Derby 1954 73C 10/72 P 101
to BR Derby Works, November 1972; overhauled and modified; exported (aboard the MV Avafors) from Middlesbrough Docks to Lamco Mining Co, Tokadeh Mine, Nimba, Liberia, May 1974; spare locomotive by 1986; last used 23rd February 1987; extant, but observed derelict, December 2011; subsequent history not known.

D3094 Derby 1954 73F 10/72 P 102
to BR Derby Works, March 1973; overhauled and modified; exported (aboard the MV Avafors) from Middlesbrough Docks to Lamco Mining Co, Tokadeh Mine, Nimba, Liberia, May 1974; last used 14th June 1980; used for spares, 1986; extant, but observed derelict, December 2011; subsequent history not known.

D3098 Derby 1955 73F 10/72 P 103
to BR Derby Works, March 1973; overhauled and modified; exported (aboard the MV Avafors) from Middlesbrough Docks to Lamco Mining Co, Tokadeh Mine, Nimba, Liberia, May 1974; last used pre-1980; used for spares, 1986; extant, but observed derelict, December 2011; subsequent history not known.

D3100 Derby 1955 75C 10/72 P 104
to BR Derby Works, March 1973; overhauled and modified; exported (aboard MV Avafors) from Middlesbrough Docks to Lamco Mining Co, Tokadeh Mine, Nimba, Liberia, May 1974; last used 14th March 1982; used for spares, 1986; extant, but observed derelict, December 2011; subsequent history not known.

Mystery locomotive: an unidentified D31XX Class 10 locomotive was photographed at Lissone, Lombardy, Italy, on 29th June 1973, and at Albate Camerlata in 1974, when owned by Attilio Rossi. It was being used on the relaying of the Milano to Como line. It was later observed at San Martino della Battaglia, Lombardy, on 1st January 1990.

D3639 Darlington 1958 36A 7/69 F ?
to C.F. Booth Ltd, Doncaster; to Conakry, Guinea, West Africa; exported from Surrey Dock, London, March 1970; used on the construction of the 135km Chemin de fer de Boke railway line, to take bauxite from mine to sea; dismantled by 1976; scrapped, early 1980s.

D3649 Darlington 1959 36A 7/69 F ?

to C.F. Booth Ltd, Doncaster; to Conakry, Guinea, West Africa; exported from Surrey Dock, London, March 1970; used on the construction of the 135km Chemin de fer de Boke railway line, to take bauxite from mine to sea; dismantled by 1976; scrapped, early 1980s.

D9505 Swindon 1964 50B 4/68 F 98 88 202 2202-4

(from main listing): to Harwich Docks, 10th July 1975; exported to Sobermai, Maldegem, near Bruges, Belgium; overhauled by Sobermai; later sold to Suikergroep N.V. Opperstraat, Moerbeke-Waas sugar factory, near Gent; working on 22nd June 1997; scrapped on site, 1999.

D9515 Swindon 1964 50B 4/68 F ?

(from main listing): to Spain; exported from Goole Docks to Bilbao, circa 16th June 1982; later stored at Chamartin Yard, Madrid, by February 1986, and still there 2nd April 1988; subsequent history not known; believed scrapped by October 2002.

D9534 Swindon 1965 50B 4/68 F ?

(from main listing): to Harwich Docks, 10th July 1975; exported to Sobermai, Maldegem, near Bruges, Belgium; overhauled by Sobermai; reported sold to an industrial user near Milan, Italy, 1976; this may be Ambrogio Transporti SPA, Gallarate, Lombardy, near Milan; reported working at a steelworks near Brescia, by May 1997; subsequent history not known; believed scrapped.

D9548 Swindon 1965 50B 4/68 F 937113106028

(from main listing): to Spain; exported from Goole Docks to Bilbao, June 1982; later stored at Chamartin Yard, Madrid, and still there January 1988; passed to infrastructure company Curbiertas and MZOV, where it was numbered P-602-03911-002-CMZ; following a merger in 1997, the owner became NECSO Entrecanales Cubiertas S.A.; seen at Sagrera goods depot, Barcelona, 19th June 1998, with NECSO EC name painted on its side and now numbered 937113106028; scrapped by October 2002.

D9549 Swindon 1965 50B 4/68 F P-601-03911-003-CMZ

(from main listing): to Spain; exported from Goole Docks to Bilbao, June 1982; later stored at Chamartin Yard, Madrid, and still there January 1988; passed to infrastructure company Curbiertas and MZOV; following a merger in 1997, the owner became NESCO Entrecanales Cubiertas S.A. and it was numbered P-601-03911-003-CMZ; to Industrias Lopez Soriano, Calle de Miguel Servet, Zaragosa, by circa 2000; extant 7th May 2003; scrapped March 2007.

- RH 221615 1943 MQ ? F ?

2ft 0in gauge locomotive of the maker's class 48DL - for details see section 26; used by Southern Railway & British Railways at Meldon Quarry, Devon; exported to Egypt, date not known; seen derelict at Nag Hammadi Sugar Factory, Egypt, 1982; the Nag Hammadi Sugar factory was built in 1895-1897 by French contractors Cail and Fives and is still in operation in 2019, but the loco's subsequent history is not known; believed scrapped.

—- RH 202005 1940 HHC ? F ?

2ft 3in gauge locomotive of the maker's class 20DL – for details see section 28; used by British Railways on an internal system around the timber stockpiles at its Sleeper Depot at Hall Hills, close to Boston Docks; locomotive acquired new in 1940; to John S. Allen & Son Ltd, Mardyke Works, Cranham, near Upminster, Essex (via Rundle & John Philips & Co Ltd), by May 1967; exported to Singapore, date not known; subsequent history not known; believed scrapped.

postscript: at the time of writing there are nine ex-BR locomotives whose disposal is unknown or not proven, but which remain suspects as having possibly been exported. These are: D2002, D2003, D2042, D2212, D2277, D2278, D2285, D2296 and D2297. If any reader has positive information as to their disposals, particularly if seen in the Queenborough area circa 1970 to 1972, the author will be pleased to receive details. In addition, during July 1969 an ex-BR diesel shunter was noted working at Imperia, northern Italy – this being in addition to the 'mystery locomotive' listed above. It has been suggested that this might be D3193, which was last seen at BR Derby Works in August 1967.

D2036, by then out of use and standing off the tracks, is seen at Siderurgica SPA, Montirone, Brescia, Italy, on 31st August 1996. (Mark Jones)

APPENDIX D : EX-LMS LOCOMOTIVES

In addition to the ex-BR locomotives in this book, there have also been a number of ex-LMS shunters which fall into the categories covered herein. Although these locomotives are not ex-BR shunters, they are closely related forerunners, and the five known *extant* examples are included here for the sake of interest and historical record. Other examples saw industrial service before being scrapped, whilst others went abroad in World War 2, some surviving to have post-war careers before being scrapped.

LMS 7050 : built for the LMS by Drewry (works number 2047 / EE 874 of 1934); delivered as LMS 7050 and spent six years dock shunting at Salford; to Air Ministry, on loan, 1940; withdrawn from LMS stock, March 1943; sold to War Department and re-numbered 224; subsequent re-numberings by the WD, and later the Army, saw it carry 70224 (in 1944), 846 (1952) and 240 (1968); later worked at Royal Navy, Botley, Hampshire; acquired for preservation in 1979; displayed at Museum of Army Transport, Beverley; to National Railway Museum, York, 2003; still at NRM in 2019.

LMS 7051 : built by the Hunslet Engine Company of Leeds (works number 1697 of 1932) as a demonstrator; underwent trials at a colliery and on the LMS; purchased by the LMS, May 1933; loaned to the War Department and numbered 27, August 1940; returned to LMS, 1941 to 1944; returned to War Department and re-numbered 70027; at the end of World War 2 it was returned to the LMS; withdrawn, December 1945; purchased by Hunslet Engine Company and used as a works shunter; it was also hired out, including time on British Railways; rebuilt by Hunslet Engine Company in 1949; preserved by Middleton Railway, Leeds, 1960; still at Middleton Railway in 2019.

LMS 7069 : built for the LMS by Hawthorn/Leslie (works number 3841 of 1935); to War Department 1940; dispatched to France as WD18 with British Expeditionary Force, 1st May 1940; abandoned at Nantes in 1940; taken over by German forces until recaptured near Le Mans in 1945; re-used by British Army at Aubigne-Racan; to Chemin de Fer et St. Calais, a privately run public railway near Le Mans, as No.7; railway closed in 1977; to L. Patry, Paris (equipment dealer) in 1978 and OOU in their yard until December 1987; repatriated to Swanage Railway, Dorset, 27th November 1987; to Blue Circle Cement Terminal, Hamworthy, about 5th April 1991; to East Lancashire Railway, Bury, about December 1994; to Gloucestershire Warwickshire Railway, Toddington, 18th April 1998; to Vale of Berkeley Railway, Sharpness, 1st September 2015; still at Sharpness in 2019.

LMS 7103 : built at Derby in 1941 and withdrawn by the LMS in December 1942 and sold to the War Department; used in Egypt; to Italy, April 1945; sold to Italian Railways (FS) in 1946; withdrawn by FS in 1984 as number 700.001; sold to a scrap merchant in Arquata Scrivia, near Genova; resold to Cariboni SPA, Colico, Italy; to Vercelli, Italy, by May 1995; to Museo Ferroviario Piemontese Store, Torino (Turin) Ponte Mosca Station, Italy, 1998.

LMS 7106 : built at Derby in 1941 and withdrawn by the LMS in December 1942 and sold to the War Department; used in North Africa in 1943, in Tunis by June 1943, in Algeria from November 1943 to March 1944; to Italy, March 1944; sold to Italian Railways (FS), 1946; numbered FS 700.003; withdrawn by FS in 1984; sold to a scrap merchant in Arquata Scrivia, near Genova; resold to Transporto Ferroviario Toscano, Arezzo Pescaiola, Italy (who operate the branch line from Arezzo, west of Florence), 1991.

APPENDIX E : BOGUS 'EX-BR' LOCOMOTIVES

Over the years various preserved industrial diesel shunters have been given running numbers which purport to be BR numbers. Sometimes these are genuine BR numbers which were once carried by now-scrapped locomotives. Sometimes they are similar to BR numbers. Sometimes locomotives carry an authentic BR livery. Such locomotives are theoretically not relevant to this book because they are not genuine ex-BR machines. However, the existence of such bogus locomotives at preservation sites can be confusing and/or misleading to visiting enthusiasts and could also prove to be a minefield for future historians. It has been decided, therefore, to provide a basic list of known examples of bogus locomotives (as a matter of historical record), so explaining why what may appear to be ex-BR locomotives at preservation sites are not in this book's main listings. No attempt has (or will) be made to provide detailed histories of these locomotives. A simple list of known relevant numbers is considered to be all that is necessary.

11230, 11509, 11510, 12139, 15097, 15099, D2447, D2700, D2870, D2911, D2957, D2959, D2960, D2961, D2971, D2999, DS1169, DS1174.
In addition 27414 is in BR livery.

Below: The Yorkshire Engine Company of Sheffield supplied twenty of its 0-4-0 diesel-hydraulic shunters to British Railways (D2850 to D2869) of which no less than eleven subsequently worked in industry, as detailed on pages 32 to 34 of this book. The item below is a reproduction of an official Yorkshire Engine Company proof for an advert (YE 47) that appeared in trade journals of the era. It depicts D2856 which later worked in industry for Redland Roadstone, as detailed on page 32. This item is from the collection of Adrian Booth.

YORKSHIRE

Yorkshire Engine 170 Diesel Hydraulic Locomotive 0-4-0 at British Railways Motive Power Depot, Bootle, Liverpool.

'YORKSHIRE' KEEPS THINGS MOVING

Yorkshire Diesel shunting locomotives are at work taking sharp curves and steep gradients in their powerful stride. These 170 h.p. diesel-hydraulic locomotives are powered by Rolls-Royce engines with torque converter, and the controls are both simple and efficient. Westinghouse vacuum controlled brakes and deadman's equipment are fitted. Cabs are designed with straight sides so that the driver has an all-round view without opening doors and windows—a great advantage in winter. 20 of these locomotives have now replaced the old Lancashire and Yorkshire Railways 'Pugs'.

A subsidiary of The United Steel Companies Limited

Powered by ROLLS-ROYCE Diesels

MEADOW HALL WORKS · SHEFFIELD 9

YE 47

01. D2066 (03066) working at Barrow Hill Engine Shed, Staveley, on 5th April 2017.
(Adrian Booth)

02. D2133 standing in the yard at West Somerset Railway, Minehead, on 14th July 2012.
(Brian Cuttell)

03. D2205 (with D2337 and D2284) preserved at Peak Rail, Rowsley, on 3rd September 2017. (Brian Cuttell)

04. D2207 at the North Yorkshire Moors Railway's C&W Works, Pickering, on 16th May 2009. (Adrian Booth)

05. D2289 newly arrived from Italy, at Peak Rail, Rowsley, on 2nd September 2018. (Brian Cuttell)

06. D2337 preserved at Peak Rail, Rowsley, on 3rd September 2017. (Brian Cuttell)

07. D2420 (06003) preserved at Peak Rail, Rowsley, on 25th June 2013.
(Brian Cuttell)

08. D2587 preserved at Peak Rail, Rowsley, on 2nd September 2018. (Brian Cuttell)

09. D2853 preserved at Barrow Hill Engine Shed, Staveley, on 5th May 2016.
(Adrian Booth)

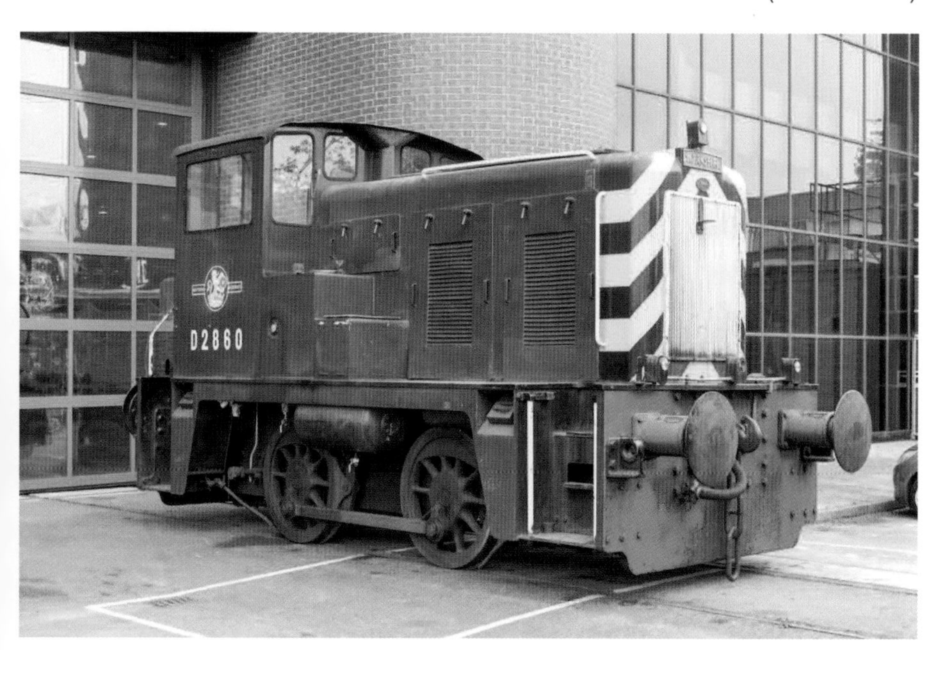

10. D2860 standing outside the National Railway Museum, York, on 8th June 2016.
(Adrian Booth)

11. D2953 preserved at Peak Rail, Rowsley, on 2nd September 2018. (Brian Cuttell)

12. D2985 (07001) preserved at Peak Rail, Rowsley, on 2nd September 2018. (Brian Cuttell)

13. D2996 (07012) preserved at Barrow Hill Engine Shed, Staveley, on 5th May 2016.
(Adrian Booth)

14. D3265 (13265, 08195) preserved at the Llangollen Railway, on 15th September 2006.
(Adrian Booth)

15. D3723 (08556) working at the North Yorkshire Moors Railway's New Bridge Yard, on 15th May 2005. (Adrian Booth)

16. D3765 (08598) preserved at the Chasewater Railway, on 16th April 2016. (Robert Pritchard)

17. D3836 (08669) working at the Wabtec Works, Doncaster, on 11th March 2015. (Adrian Booth)

18. D3898 (08730) awaiting tyre turning at Midland Road Depot, Leeds, on 1st February 2017. (Adrian Booth)

19. D3986 (08818, 4, MOLLY) at Barrow Hill Engine Shed, Staveley, on 18th July 2017. (Adrian Booth)

20. D4018 (08850) outside the shed at North Yorkshire Moors Railway, Grosmont, on 10th April 2004. (Adrian Booth)

21. D4092 preserved at Barrow Hill Engine Shed, Staveley, on 1st May 2017.
(Brian Cuttell)

22. D4106 (09018) on-hire at Hope Cement Works, Derbyshire, on 15th February 2013.
(Adrian Booth)

23. D3536 (09201) in the shed yard at Hope Cement Works, Derbyshire, on 10th May 2017. (Robert Pritchard)

24. D9514 working at Ashington Colliery, Northumberland, on 31st May 1979. (Robert Pritchard)

25. D9539 preserved at Peak Rail, Rowsley, on 28th May 2016. (Brian Cuttell)

26. 15224 in the shed yard at Snowdown Colliery, Kent, on 8th September 1980. (Robert Pritchard)

27. ZM32 preserved at the Steeple Grange Light Railway, Derbyshire, on 29th August 2010. (Brian Cuttell)

28. D226 in the shed yard at Keighley & Worth Valley Railway, Haworth, on 13th May 2004. (Adrian Booth)

29. Ruston & Hornsby 224337 of 1944 at J. & K. Harris's yard, March, on 21st April 1989. (Andrew Smith)

30. PWM654 preserved at Peak Rail, Rowsley, on 2nd September 2018. (Brian Cuttell)

31. 7106 (ex-LMS) at Calbenzano, Italy, on 12th October 2017. (D. Pollock)

32. Bogus D2999 (Brush 91 of 1958) at Middleton Railway, Leeds, on 24th August 2003. (Adrian Booth)